What Executives
Need to Know about
PROJECT
MANAGEMENT

What Executives Need to

Know about

PROJECT

MANAGEMENT

Harold Kerzner, Ph.D.

Frank P. Saladis, PMP

WILEY

John Wiley & Sons, Inc.

INTERNATIONAL
Institute for Learning, Inc.

This book is printed on acid-free paper. ∞

Copyright © 2009 by International Institute for Learning, Inc., New York, New York.
All rights reserved.

Published by John Wiley & Sons, Inc., Hoboken, New Jersey

Published simultaneously in Canada

Photo credits-- Figures 2.1, 2.2, 2.3, 3.2, 3.3, 3.4, 3.5, 3.6, 3.7, 3.9, 3.10, 4.2, 4.3, 5.5, 5.8, 5.10,
5.15, 5.16, 5.17, 5.18, 5.21, 5.23, 5.28, 5.49, 5.52, 5.54, 6.1, 7.8, 7.13, 7.14--PhotoDisc/Getty
Images; Figures 2.6, 2.8, 2.10, 3.8, 5.2, 5.7, 5.9, 5.19, 5.20, 5.24, 5.27, 5.30, 5.35, 5.47, 5.51,
7.9, 7.12, 7.15--Digital Vision; Figures 3.1, 5.12, 5.29, 6.12, 7.11--Artville/Getty Images; Figures
3.11, 3.12, 4.1, 5.31, 5.55, 7.7--Corbis Digital Stock; Figures 5.4, 5.53--ImageState; Figures 5.6,
5.11, 5.22, 5.32, 5.48, 5.50, 5.56, 6.2, 6.3, 6.7, 6.10, 6.11, 8.1, 8.2, 8.3, 8.4--Purestock; Figures
5.33--StockByte/Getty Images; Figures 7.10--IT Stock

For general information about our other products and services, please contact our Customer
Care Department within the United States at (800) 762-2974, outside the United States at
(317) 572-3993 or fax (317) 572-4002.

Wiley also publishes its books in a variety of electronic formats. Some content that appears in
print may not be available in electronic books. For more information about Wiley products, visit
our web site at www.wiley.com.

"PMI", the PMI logo, "OPM3", "PMP", "PMBOK" are registered marks of Project Management
Institute, Inc. For a comprehensive list of PMI marks, contact the PMI Legal Department.

Library of Congress Cataloging-in-Publication Data:
Kerzner, Harold.
      What executives need to know about project management / Harold Kerzner, Frank Saladis.
         p.   cm.—(The IIL/Wiley series in project management)
      Includes index.
      ISBN 978-0-470-50081-1 (cloth)
         1. Project management.   2. Management.   I. Saladis, Frank P.   II. Title.
      HD69.P75K497   2009
      658.4'04—dc22
                                                                      2009018605

Printed in the United States of America

10 9 8 7 6 5 4 3 2 1

CONTENTS

Chapter 7:
NEW CHALLENGES FACING SENIOR MANAGEMENT   227

Chapter 8:

PREFACE

In the early years of what we refer to today as *modern project management,* executives had apprehensions about accepting changes to what they believed was an effective approach to organizational management and leadership. The apprehensions revolved mainly around how much authority should be delegated to project managers and whether project managers would then be asked to make decisions that were normally decisions reserved for senior management. A primary customer of the outputs of formal project management, the Department of Defense, wanted project managers to be capable of answering any and all questions about a project, especially technical questions. This mandated that the project managers possess a solid and reliable command of technology. Engineers with advanced degrees and a reputation as technical experts were assigned as project managers. The decision to utilize technical experts created some concern and even made some executives fearful that the highly technical project managers would make decisions without first consulting with senior management.

Project management was treated by executives in many organizations as a part-time assignment and not an actual occupation. There was no organization support structure and methodology, and at the completion of a project, the project manager would return to his or her previous line or operations function. Executives tried to manage most aspects of the project from the executive levels and treated the project managers more or less as coordinators or administrative resources with very little formal authority. But all of this was about to change.

Military officers who awarded the contracts and had the responsibility for monitoring the performance of the contractors after the contract go-ahead decision considered their hierarchical equals in the contractor's firm to be the executives, not the project managers. Executives now had to be more visibly involved in the projects, and thus the role of the executive project sponsor appeared. The question then becomes, "Would the role of the executive as a sponsor be more than just executive-client contact?"

The majority of the contracts were yearly renewable-type contracts, and project management was beginning to emerge as a profession. The first challenge facing the executives was to determine whether the project manager position should be placed on the management career path ladder, the technical career ladder, or an entirely new career ladder entitled Project Management. The eventual series of decisions resulted in the creation of a project management as a career path with its own ladder of professional development.

Most of the engineers who were assigned as project managers had strong technical skills and a command of the technology involved, but had a limited knowledge of the business and, in some cases, even the industry they were in. The project sponsor or executive had the responsibility to make sure that the project team and the project manager understood the business need for the project, the company's objectives in accepting the contract, and the business risks associated with the project. The engineers also had a reputation for being highly confident and optimistic in their planning and often did not see any reason for developing contingency plans. It was necessary for the sponsor to be actively involved in the initiation phase of the project to make sure that planning was done correctly and in compliance with the customer's requirements and to meet organizational goals.

The engineering project managers were highly trained in technology but generally lacked training in human resource–type skills, which we commonly refer to as "soft skills" or interpersonal skills. When behavioral issues and interpersonal conflicts occurred during project implementation, the project sponsor had to step in and resolve the problems.

The Department of Defense mandated that project management methods should be used on projects that exceeded a certain dollar threshold limit. Contractors began developing project management methodologies designed for projects that exceeded the threshold limit. It soon became apparent that project management, as a methodology, could be used effectively on any and all projects, not merely those that exceeded the threshold limit. Sponsors were now required for all projects, and templates describing the role of the project sponsors became a part of the methodology. Several types of project sponsorships were required to manage the needs of the organization, and templates for these sponsorships were created. The main reason for the variety of sponsor types was the fact that executives could not act as sponsors for every project. The demands of the organization were too great and required their attention in many other areas. It was determined that some projects could be overseen by sponsors from lower and middle levels of management, depending on the type of project and its relationship to and impact on the organization.

As project management matured, so did the executive's role in project management. Executives became actively involved in activities such as capacity planning for projects, portfolio management, prioritization, and strategic planning for project management. Many executives today are certified Project Management Professionals® (PMP®s) and identify themselves as such in proposals during competitive bidding activities. Executives in today's business environment are now active participants and visible supporters of project management as a factor in organizational success.

Harold Kerzner, Ph.D.

Frank P. Saladis, PMP

International Institute for Learning, Inc., 2009

ACKNOWLEDGMENTS

Some of the material in this book has been either extracted or adapted from Harold Kerzner's *Project Management: A Systems Approach to Planning, Scheduling, and Controlling*, 10th edition; *Advanced Project Management: Best Practices on Implementation*, 2nd edition; *Strategic Planning for Project Management Using a Project Management Maturity Model; Project Management Best Practices: Achieving Global Excellence*, 1st edition (all published by John Wiley & Sons, Inc.).

Reproduced by permission of Harold Kerzner and John Wiley & Sons, Inc.

We would like to sincerely thank the dedicated people assigned to this project, especially the International Institute for Learning, Inc. (IIL) staff and John Wiley staff for their patience, professionalism, and guidance during the development of this book.

We would also like to thank E. LaVerne Johnson, Founder, President & CEO, IIL, for her vision and continued support of the project management profession, Judith W. Umlas, Senior Vice President, Learning Innovations, IIL and John Kenneth White, MA, PMP, Senior Consultant, IIL for their diligence and valuable insight.

In addition, we would like to acknowledge the many project managers whose ideas, thoughts, and observations inspired us to initiate this project.

HAROLD KERZNER, PH.D., AND FRANK P. SALADIS, PMP

INTERNATIONAL INSTITUTE FOR LEARNING, INC. (IIL)

International Institute for Learning, Inc. (IIL) specializes in professional training and comprehensive consulting services that improve the effectiveness and productivity of individuals and organizations.

As a recognized global leader, IIL offers comprehensive learning solutions in hard and soft skills for individuals, as well as training in enterprise-wide Project, Program, and Portfolio Management; PRINCE2®*; Lean Six Sigma; Microsoft® Office Project and Project Server**; and Business Analysis.

After you have completed *What Executives Need To Know About Project Management,* IIL invites you to explore our supplementary course offerings. Through an interactive, instructor-led environment, these virtual courses will provide you with even more tools and skills for delivering the value that your customers and stakeholders have come to expect.

For more information, visit www.iil.com or call 1-212-758-0177.

---

*PRINCE2® is a registered trademark of the Office of Government Commerce in the United Kingdom and other countries.
**Microsoft Office Project and Microsoft Office Project Server are registered trademarks of the Microsoft Corporation.

Chapter

PROJECT MANAGEMENT PRINCIPLES

THE TRIPLE CONSTRAINT

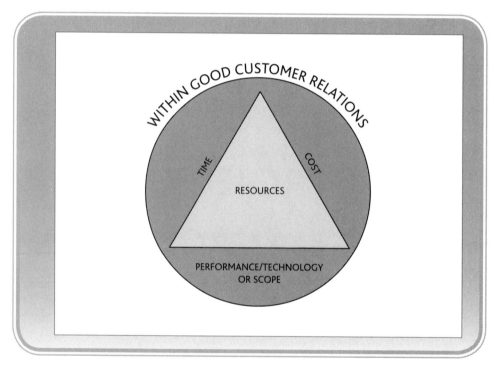

Effective project management is an attempt to improve the efficiency and effectiveness of an organization by arranging for work to flow multidirectionally through the organization. Project management was developed to focus on organizational activities that had the following characteristics:

- Unique or one-of-a-kind deliverables

- A well-defined objective

- Predetermined constraints regarding time, cost, and performance/technology/quality

- Requires the use of human and nonhuman resources

- Has a multidirectional work flow

From an executive perspective, the figure illustrates the basic goal of project management, namely, meeting the objectives associated with the triple constraint of time, cost, and performance while maintaining good customer relations. Unfortunately, because most projects have some unique characteristics, highly accurate estimating may not be possible and trade-offs between the triple constraint may be necessary. Executive management must be involved in almost all of the trade-off discussions to make sure that the final decision is made in the best interest of both the project and the company. Project managers may possess sufficient technical knowledge to deal with many day-to-day decisions regarding project performance but may not have sufficient business knowledge to adequately address and care for the higher-level, broader interests of the company.

TYPES OF PROJECT RESOURCES

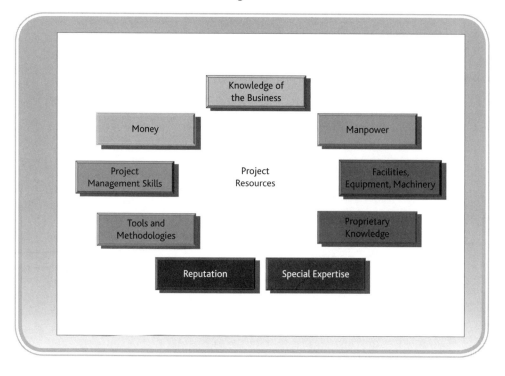

This illustration shows the various project resources that project managers may or may not have under their direct control. Some of these resources require additional comment.

- *Money.* Once budgets are established and charge numbers are opened, project managers focus more on project monitoring of the budget rather than management of the budget. Once the charge numbers are approved and opened, the respective line managers or functional managers control the budgets for each work package.

- *Resources.* The human resources required for the project are usually assigned by the line managers, and these resources may be under the direct control of the line managers for the duration of the project. Also, even though the employees are assigned to a project team, their line managers may not authorize them to make decisions that affect the functional group without first obtaining approval from the line managers.

- *Business knowledge.* Project managers are expected to make decisions that will benefit the business as well as the project. This is why executives must interface with projects—to provide project managers with the necessary business information for decision making.

THE EVOLUTION OF PROJECT MANAGEMENT

EVOLUTION

Over the past five decades, there have been rapid evolutionary changes in the way projects are managed. For simplicity's sake, they will be broken down into three categories as follows:

- *1960–1985.* This can be referred to as the period of traditional project management. Project management was restricted to aerospace, defense, and heavy construction industries. Project management was used on mega projects only.

- *1986–1992.* This was the Renaissance period, or great awakening, where we learned that a project management methodology could be used on a multitude of projects and benefit nearly all industries. Project management became readily accepted in industries such as automotive, information systems, telecommunications, and banking.

- *1993–2009.* Project management became readily accepted in all industries and applicable to any size project. Project management was viewed as a career path, and interest in project management certification credentials grew. Companies began to recognize that project management can increase profitability and improve working relationships with customers while increasing competitive edge.

PROJECT OBJECTIVES

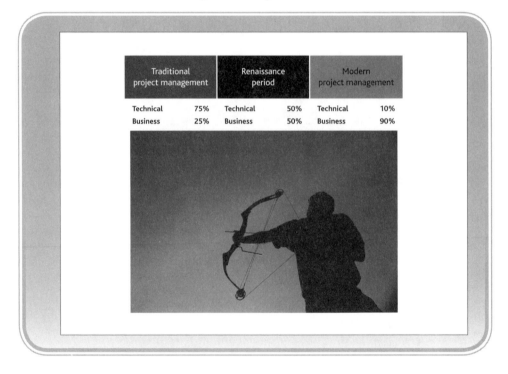

Traditional project management		Renaissance period		Modern project management	
Technical	75%	Technical	50%	Technical	10%
Business	25%	Business	50%	Business	90%

Over the years, project management has matured into a business process as well as an organized product and service delivery process. As such, project objectives are now more closely aligned with business objectives in addition to technical objectives.

Historically, this was due to the fact that people with engineering backgrounds were placed in charge of projects. These people had a command of technology but did not generally understand or were not required to develop a working knowledge about the business. The ultimate objective was to produce a high-technology deliverable regardless of the cost or time required. As long as the deliverable worked, the project was considered to be a success.

Today's objectives focus more on the necessity to satisfy a business objective rather than merely a technical objective. Executive involvement in establishing business objectives is essential. Many project managers have limited business knowledge (but excellent technical skills) and must rely heavily on senior management to communicate business-related assumptions and constraints. Senior management's involvement is also necessary to make sure that the project objectives and the business objectives of the project are aligned with the company's strategic objectives. Developing technology just for technology's sake is simply not useful for any business.

DEFINITION OF *SUCCESS*

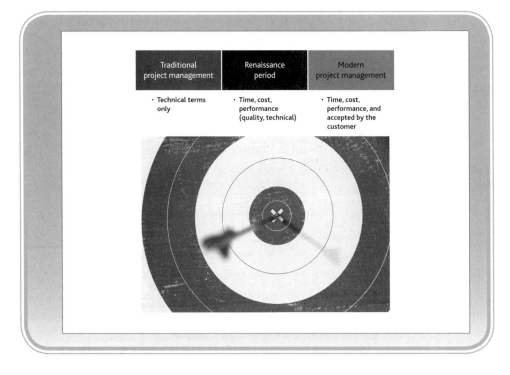

Over the years, our definition of *success* has changed from technical success criteria (functionality) to business-related success criteria—specifically, on time, within planned cost, at the desired performance level, and with formal customer acceptance. Today, success for each project undertaken is now defined through a mutual agreement between the customer and the contractor or performing organization. Other definitions of success include:

- Additional business with the client

- The client acts as an advocate for your company

- Increased market share

- Shorter time to market

The definition of project success must be decided upon during the initial discussions and negotiations with the customer. The agreed-upon definition of success must be in alignment with organizational goals, and senior management should clearly and visibly support the agreed-upon success criteria.

VELOCITY OF CHANGE

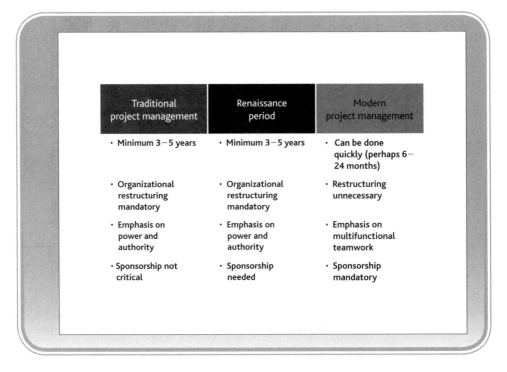

Traditional project management	Renaissance period	Modern project management
· Minimum 3−5 years	· Minimum 3−5 years	· Can be done quickly (perhaps 6−24 months)
· Organizational restructuring mandatory	· Organizational restructuring mandatory	· Restructuring unnecessary
· Emphasis on power and authority	· Emphasis on power and authority	· Emphasis on multifunctional teamwork
· Sponsorship not critical	· Sponsorship needed	· Sponsorship mandatory

Historically, organizational restructuring became a necessity with the award of each large contract. If the project were large enough, a separate organizational unit would be created just to service this one project. And with each organizational change, some executives increased their level of authority, while others may have experienced a reduction of their authority and power. The profitability of each project had a major impact on the bonus for the executive in control of that project. Each executive would establish a separate and unique culture for his or her project and, when the project was completed, the culture was expected to dissolve. Unfortunately, large companies with a multitude of projects in progress also had several cultures in place at the same time, and many of these cultures experienced conflicts with other cultures. As a result, many executives, especially those who did not have direct involvement with projects, became disenchanted with project management.

As project management evolved, it became apparent that the secret to effective project management was effective communications, cooperation, teamwork, and trust, rather than power and authority. It was determined that project management could work effectively within any organizational structure, and restructuring was deemed unnecessary. Most executives were pleased with this result.

AUTHORITY AND JOB DESCRIPTIONS

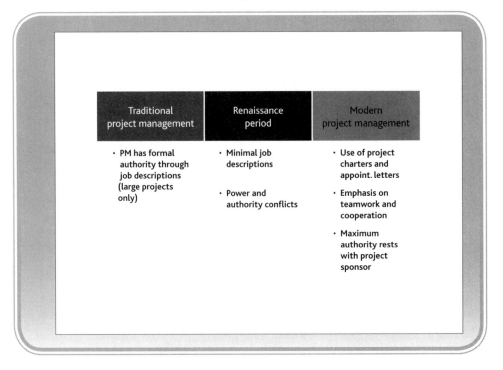

Traditional project management	Renaissance period	Modern project management
· PM has formal authority through job descriptions (large projects only)	· Minimal job descriptions · Power and authority conflicts	· Use of project charters and appoint. letters · Emphasis on teamwork and cooperation · Maximum authority rests with project sponsor

One of the greatest concerns that executives struggled with over the years was deciding about how much authority should be given to the project managers. Executives were not willing to relinquish any part of their hard-earned authority to the project managers for decision making. Some companies reluctantly decided to make project management a career path position but provided the project manager with limited authority. The fear was that if project managers played a greater part in the decision process, they would eventually share in the distribution of year-end bonuses.

In today's project environment, maximum authority generally resides with the project sponsor, and whatever authority is granted to the project manager is defined in the project charter. Although some formal authority is important for the project manager position, managers and executives have realized that project management works well when the focus is on teamwork, communication, cooperation, trust, and the project objectives rather than on formal authority.

Project managers may have more implied authority than real authority. Executives expect the project managers to assume as much authority as needed to get the job done while minimizing conflicts in the process. Using the project charter as a means of documenting and communicating the project manager's authority is a common method for defining authority on a project-by-project basis.

EVALUATION OF TEAM MEMBERS

Project managers historically were not provided with any wage and salary administration responsibility. There were several reasons for this:

- It was common practice for project managers to have team members that were at a higher pay grade than the project manager.

- Team members did not interface directly with the project manager, and the project manager did not know them well enough to administer an evaluation.

- Team members worked on multiple projects concurrently, and each project manager could have a different interpretation as to how well or how poorly the worker performed.

- Project managers were not trained in wage and salary administration.

Today, project managers are allowed to provide input to the evaluation process, but the final decision about performance is made by the line or functional manager. The input is usually based on the project manager's observation of the worker's communication skills, cooperation, teamwork, planning skills, and execution skills. However, unless the project manager is a technical expert, he or she may not be qualified to evaluate the worker's technical performance.

ACCOUNTABILITY

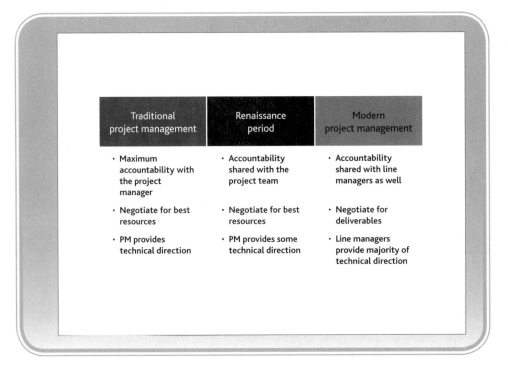

Traditional project management	Renaissance period	Modern project management
• Maximum accountability with the project manager	• Accountability shared with the project team	• Accountability shared with line managers as well
• Negotiate for best resources	• Negotiate for best resources	• Negotiate for deliverables
• PM provides technical direction	• PM provides some technical direction	• Line managers provide majority of technical direction

Historically, project managers were selected from the engineering ranks and were expected to possess a command of technology. In some cases, team members received direction from the project managers rather than the line managers, but this was based on approval by the line managers, who determined that the project manager's technical capability was acceptable and respected the project manager. In these cases, maximum project accountability rested with the project managers.

Today, because of project complexities and rapid technology changes, project managers are expected to possess more of an understanding of technology than a command of technology. The team members may receive some or all of their direction from their line managers or subject matter experts. In many situations, project accountability has become a shared effort between the project and line managers.

Line managers are now regarded as part of the project team and are expected to share in the rewards that may be provided to the team upon successful project performance. Because of shared project accountability, line managers are now being trained in project management methods the same way project managers receive training.

PROJECT MANAGEMENT SKILLS

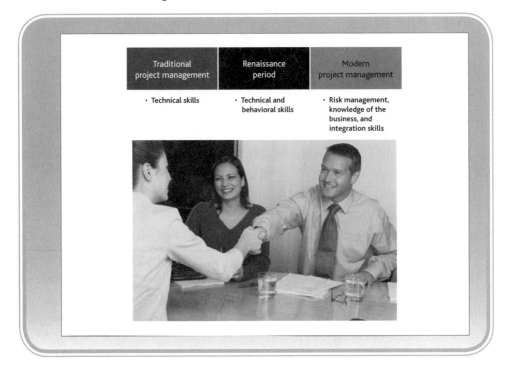

Historically, project managers were selected from the technical ranks. They possessed advanced degrees in engineering and had a very thorough knowledge of technology. Virtually all technical direction was provided by the project manager contingent upon the approval of the line managers involved.

As projects grew in complexity, project managers could no longer maintain a detailed knowledge and command of technology in all areas of the project. Also, more people with business backgrounds were being placed into project management roles. The result was that the skills required as a project manager changed from those of a technical expert to skills associated with risk management, business management, and integration of work throughout the company.

Today, project management skills are becoming more business-oriented skills. Many executives now believe that they are managing their business as though it is a series of projects, and project managers are expected to make decisions that are in the best interest of both the project and the business or company. Also, project management methodologies now include and emphasize business processes as well as project management processes.

MANAGEMENT STYLE

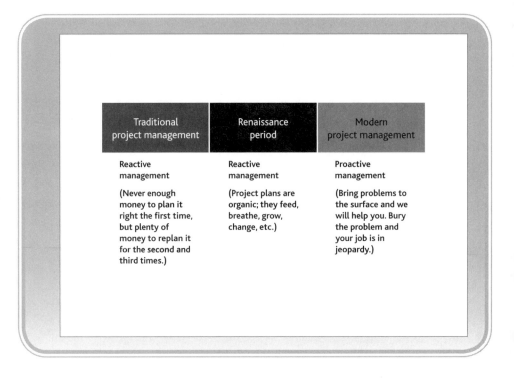

Traditional project management	Renaissance period	Modern project management
Reactive management	Reactive management	Proactive management
(Never enough money to plan it right the first time, but plenty of money to replan it for the second and third times.)	(Project plans are organic; they feed, breathe, grow, change, etc.)	(Bring problems to the surface and we will help you. Bury the problem and your job is in jeopardy.)

As stated previously, engineers with advanced degrees were frequently assigned as the project managers. These individuals were highly optimistic in general and believed that whatever plan was developed would work well and the need for contingency planning was not necessary. While this optimism may be a good trait, it came with the disadvantage that project management leadership was perceived to be based on reactive management rather than proactive management.

Today, more project managers are being selected from the business ranks rather than technical ranks. These people are more inclined to support proactive management and see the necessity for contingency planning. While changes in technology can and do occur over the life cycle of a project, changes in the business environment are more likely to occur at a much faster rate.

Without contingency plans, projects often come to a complete standstill when a serious problem occurs and people wait for new direction. Resources with critical skills may be reassigned to other projects because of the uncertainties with the existing project. Without contingency plans, replanning may encompass the entire project rather than just the crisis situation.

PROJECT SPONSORSHIP

In the early years, it became obvious that many engineers did not understand the business well enough to make sound business decisions in connection with their assigned projects. As a result, executives felt that they could not separate themselves entirely from the management of projects. The role of the executive and how to interface with a project needed to be defined. The result was the creation of an executive project sponsor position. But certain questions needed to be answered, which will be discussed later in this book:

- Can the project sponsor change over the life cycle of the project, or should the same person function as the sponsor for the entire project?

- Can sponsorship be delegated down to middle or lower levels of management in a multiproject environment?

- Can sponsorship responsibilities be managed by a committee? If so, are there limitations regarding the size of the committee?

- Must the project sponsor be assigned from the organization that is funding the project, or can the sponsor be an impartial outsider?

- What role will the sponsor assume during each life-cycle phase, and will each role mandate a different level of involvement?

PROJECT FAILURES

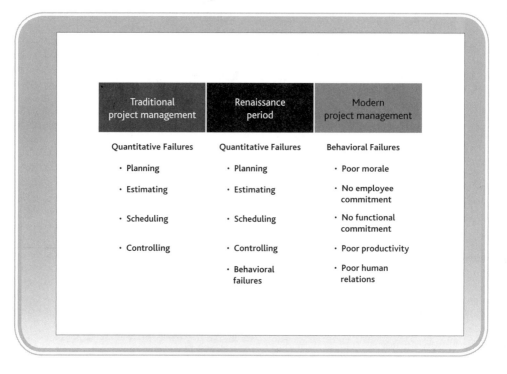

Traditional project management	Renaissance period	Modern project management
Quantitative Failures	Quantitative Failures	Behavioral Failures
• Planning	• Planning	• Poor morale
• Estimating	• Estimating	• No employee commitment
• Scheduling	• Scheduling	• No functional commitment
• Controlling	• Controlling	• Poor productivity
	• Behavioral failures	• Poor human relations

Over the years, if a project was assessed as a failure, the project managers were blamed and often relieved of their duties and responsibilities. Failures were most often identified as quantitative failures and were associated with poor planning, unreliable cost and schedule estimating, and poor control. Project managers were being held accountable for each of these even if the work was being accomplished by organizational units that were not under the direct control of the project manager.

While these are valid reasons for the failure of projects, managers often overlooked the fact that failure may have been the result of behavior issues such as poor morale, lack of employee commitment, lack of functional unit commitment or support, poor productivity, and poor human relations. The importance of the behavioral side of project management has become readily apparent today. In today's project management education environment, we now appear to have a balance between behavioral courses (soft skills) and quantitative courses (hard skills).

IMPROVEMENT OPPORTUNITIES

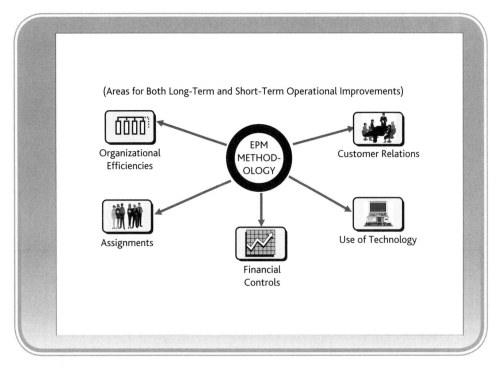

The implementation of project management can provide significant opportunities for improvement in various parts of the company:

- *Organizational efficiencies.* Processes can be developed that make organizational work flow more efficient and more effective. This can improve profit margins.

- *Customer relations.* Project management allows us to work more closely with our customers and possibly receive sole-source contracts. It also increases our chances for follow-on work from the same client.

- *Assignments.* Assigning people to project teams becomes a more efficient process, and resource capacity planning models can be developed.

- *Financial controls.* Better financial controls will be in place, both horizontally and vertically. This may necessitate the implementation of an earned value measurement system.

- *Technology.* Technology usage is observed on a company-wide basis rather than an individual department view.

RESISTANCE TO CHANGE

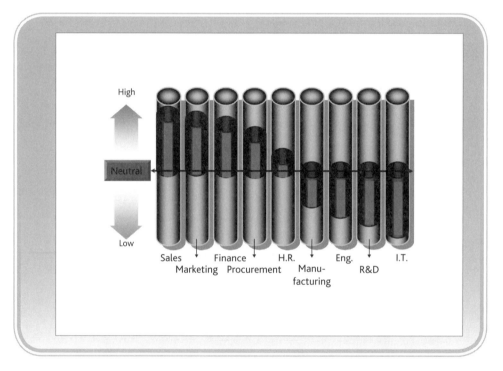

Executives must realize that each functional area will have its own view of project management, and some resistance to change can be expected. The reasons for this might be:

- Sales and marketing view project management as a risk to their year-end bonuses. They may not support project management for fear of having to share bonuses.

- Finance feels threatened by the implementation of an earned value measurement system. This will require that they learn multiple accounting systems.

- Procurement managers may be concerned about a loss of centralized control and that project managers will perform their own procurement activities.

- Human resources may be reluctant to design a whole new training curriculum for project management.

THE BENEFITS
OF PROJECT
MANAGEMENT

BENEFITS

Efficiency:

- ☐ Project management allows us to accomplish more work in less time, with fewer resources, and without any sacrifice to quality.

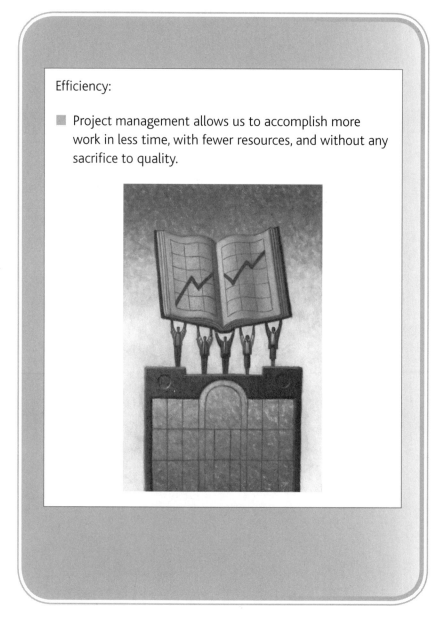

Previously, we discussed some of the benefits of implementing a project management process. The primary benefit of project management is the streamlining of the work flow, which in turn allows us to accomplish more work in less time, with fewer resources, and without any sacrifice to quality. This normally requires the implementation of a project management methodology based upon user-friendly processes, forms, guidelines, templates, and checklists.

Methodologies are developed or improved based on lessons learned and best practices that are captured at the end of each project or at the gate review meetings. The information is then used to generate continuous improvements to the project management methodology. External benchmarking activities are also good ways to generate intellectual property for use in improving the methodology.

Project management works, and works well, but as stated previously, there will be some resistance to change. Some people will put their own personal aspirations for power and authority ahead of decisions that may be in the best interest of the company. The result can be a very slow maturity process.

Profitability:

■ Under normal business conditions, profitability can be expected to increase as a result of using project management processes.

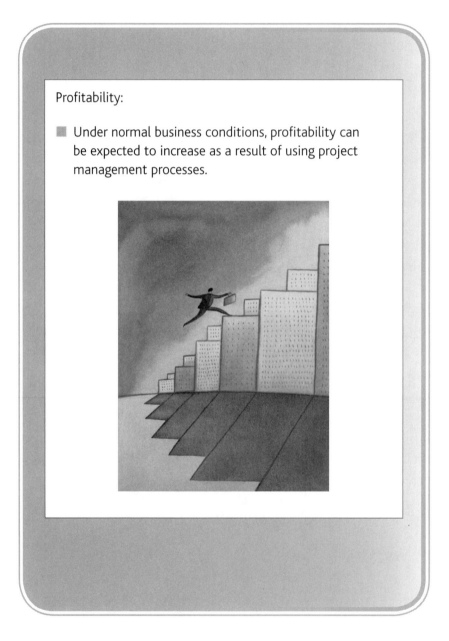

Because project management improves efficiency, profits can be expected to rise. For companies that survive on competitive bidding, profitability may very well be based on a series of streamlined processes, such as the implementation of an enterprise project management (EPM) methodology. Streamlined processes can lead to paperless project management systems that can save companies significant money.

Some customers believe that cost savings will be realized if a project manager is not assigned. In these situations, as in the case of an information technology (IT) project, the systems programmer or another functional manager will take on the additional role of the project manager. This has a great potential for failure because, without a project manager overseeing the integration of the technical components and managing from a systems level, there is significant risk that budget and schedule commitments will not be met.

Although project management can increase profitability, there are other business decisions that can increase or decrease the expected profits. Many of these other factors are defined as enterprise environmental factors and include inflation, recessions, available resources, competition, and changes in technology.

Scope Changes:

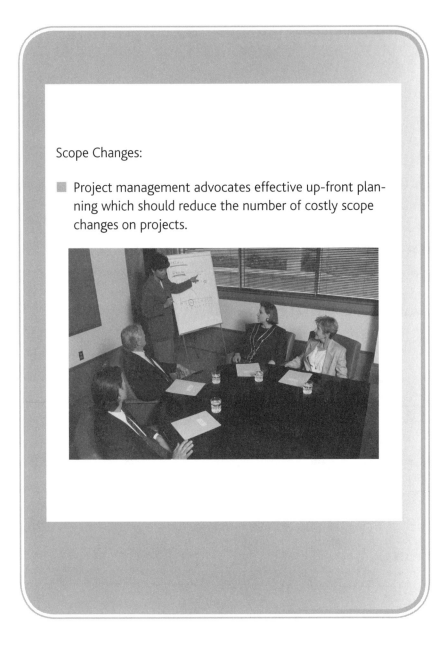

- Project management advocates effective up-front planning which should reduce the number of costly scope changes on projects.

For years, customer-funded scope changes were viewed as a source of revenue and additional profits. Project managers were pressured to push through scope changes, whether necessary or not, to generate additional income. During competitive bidding, projects were commonly underbid by 20 percent to 30 percent to win the contract with the intent that the additional profit would come from scope changes after the original contract was awarded.

Today, companies are running lean and mean, and do not have the excessive resources that were available in the past. While pushing these resources through scope changes may appear to be the right thing to do, it is the resources that are required to implement these scope changes. Limited resources are a reality for most projects, and the organization's resources are already committed to other projects. The acceptance of a scope change may require that resources be temporarily taken from other ongoing projects.

Scope changes have merit if the changes can be shown to add significant value to the customer's deliverables. Pushing through scope changes for the sake of additional profits can alienate the customer and limit follow-on work. Companies should maintain a structured change control process that involves key stakeholders with approval authority and voting rights.

Organizational Stability:

■ Project management allows work to flow in a multidirectional manner and usually results in added efficiency,
effectiveness, and stability of the organization.

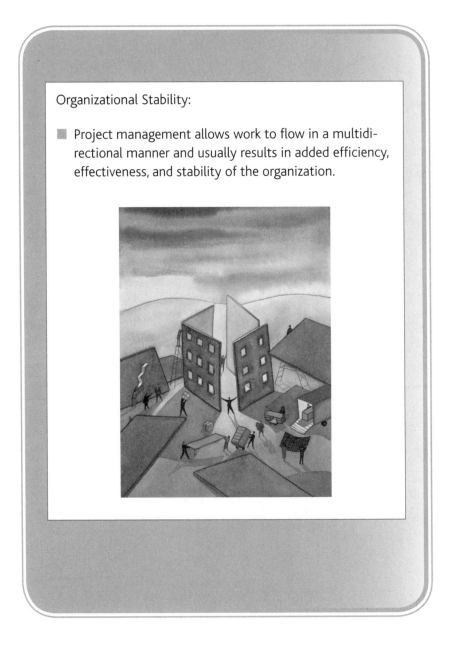

Project management allows work to flow multidirectionally, that is, horizontally and vertically at the same time. Project managers are generally given authority to talk to anyone in the company concerning the project. Decision making occurs quickly, and less time is wasted going through the often highly bureaucratic chain of command for approvals.

This results in added efficiency, effectiveness, and organizational stability but may require that executives approve this arrangement and formally provide the project managers with some degree of authority to make decisions. Organizational stability is achieved when executives make decisions based on what is in the best interest of the company before what is in their own personal interest. The organizational culture is the key here, and senior management is the architect of the culture.

Closeness to the Customers:

■ Project management allows us to work more closely with the customers, resulting in a high degree of customer satisfaction.

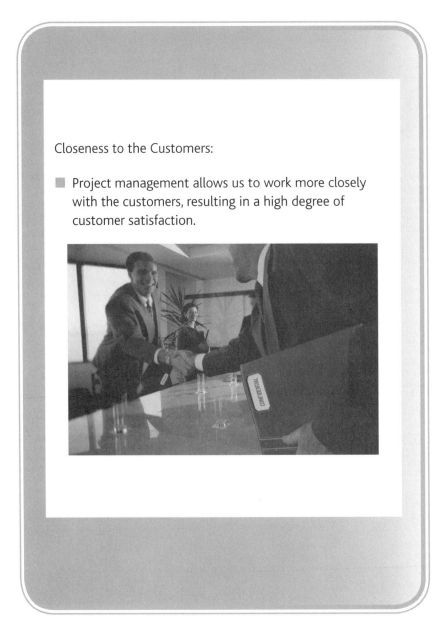

Project management allows us to work more closely with our customers, resulting in a high degree of customer satisfaction. Customer satisfaction can lead to sole-source contracting and save hundreds of thousands of dollars by not having to go through the formalized competitive bidding process.

Some companies are creating enterprise project management (EPM) methodologies that interface directly with the customer's project management methodology. After project closure, we interview the customers to seek input for improvements and recommendations for changes to our methodology that will benefit future projects for this customer. Since customers are now tracking projects through the contractor's project management methodology, contractors are providing training for customers about how the methodology works.

Problem Solving:

- Project management allows for better problem solving and usually in a shorter period of time.

Project management can create a foundation for better decision making and accelerate the decision process. But this requires that senior management provide project managers with sufficient decision-making authority. This may not occur until the organization achieves some degree of project management maturity.

The concept of integrated project teams (IPTs) has become popular during the last decade. IPTs are based on (1) selecting the appropriate team members that possess the technical know-how and (2) providing the team with the authority to make decisions. This concept of using IPTs has been shown to accelerate decision making and reduce the project's life-cycle cost and completion time. Companies that have experienced the benefits of this technique include Hewlett-Packard, 3M, and Daimler-Chrysler.

When the team possesses the technical and business knowledge along with the authority to make decisions, the team requires significantly less interfacing time with organizations external to the project. The need for communicating and interfacing with sponsors regarding decision making or seeking out subject matter experts that must ramp up to become familiar with the technical issues is minimized.

Application:

■ Project management processes can be applied to all projects, regardless of the size and scope of the projects.

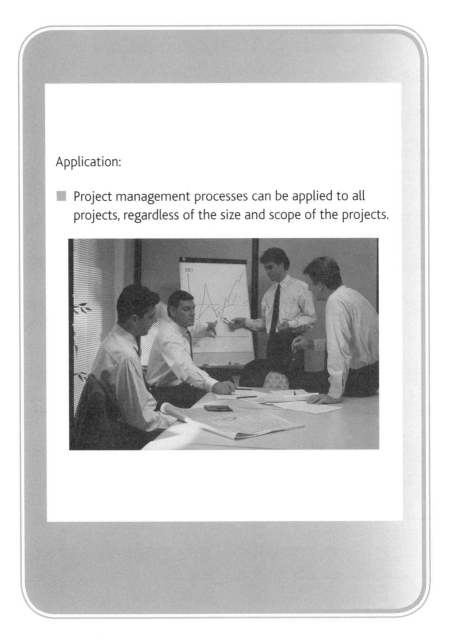

Historically, project management was used only for large projects or those above a certain threshold limit, such as:

- Exceeding a certain dollar value

- Exceeding a certain time duration

- Requiring integration across a certain number of functional units

- Special customer requirements

- Project risks

Today, we believe that companies should develop an EPM methodology. This methodology should be used for all projects regardless of the size and scope, but only those portions of the methodology that are appropriate for the project should be used. It may not be cost effective for the entire methodology to be applied to all projects.

Establishing threshold limits is usually a bad idea because it is possible for someone in a position of authority to arbitrarily change the threshold limits to avoid the use of the project management methodology. Methodologies allow problems to surface quickly rather than being hidden or buried, and also serve as early warning indicators that problems are about to surface.

Quality:

- Effective use of project management processes can increase quality without an accompanying increase in cost.

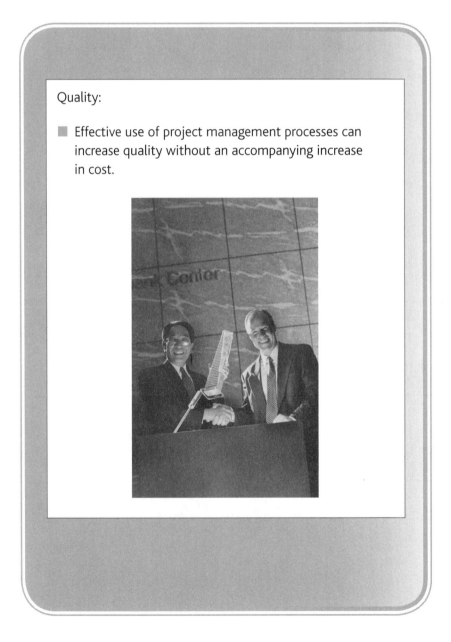

Effective project management can improve quality without any accompanying increase in cost. Most people do not fully understand the marriage between project management and quality. People seem to understand that project planning may very well be the most important life-cycle phase of a project, at least through the eyes of the project manager. But what is the most important life-cycle phase when considering quality?

Quality is *not simply* inspection. Quality must be planned for, and therefore the planning or the design phase could be regarded as the most important phase for quality. The "marriage" between project management, quality, and possibly Six Sigma strategies occurs in planning.

The EPM methodology is a structured process. Using this methodology provides some degree of structure to the implementation of quality. Six Sigma project managers must recognize the importance of using a statement of work, work breakdown structures (WBS), and schedules and understand how these planning elements may impact quality.

Authority Issues:

- Contrary to popular belief, project management actually reduces authority and power issues in companies.

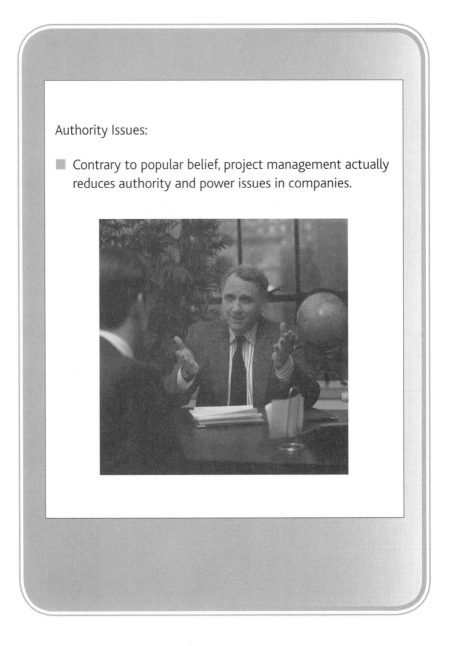

Effective project management could be defined as leadership without authority. In virtually all companies, the real authority is already determined and divided among the executives, middle management, and some lower-level managers. It is unrealistic for any project manager to believe that, simply because they have been assigned to a project, they will receive a project charter containing a vast amount of authority delegated to them by senior management.

Most project managers seem to have more implied authority than real authority. The real authority resides with senior management and the project sponsors. Project managers exercise authority through referent power. Once this is recognized and accepted, conflicts over power and authority are diminished.

However, many conflicts can be minimized or resolved quickly if whatever authority is granted to the project manager is officially documented. Employees must know what authority the project manager possesses with regard to decision making. Likewise, project managers must know how much authority the project team members have with regard to making decisions for their respective departments.

Cost of Using Project Management:

■ The cost of implementing and using project management is low, and the result should be an increase in business rather than a decrease.

Four decades ago, companies refused to consider the use of formal project management for fear that the cost of implementation would make the company noncompetitive. There was a very real fear that new layers of management would be needed to provide supervision for project management activities. This would be costly and significantly drive up the overhead costs.

Today, we realize that the supervision costs are not necessary, and project sponsorship can fulfill the supervision role. However, there may be an increase in cost initially as we put project management processes in place and create an EPM methodology. There may also be an initial cost for project management training and education as well. Within a year or two after implementation, the organization should become more efficient and more effective in the way it executes project work and the cost of using project management should be reduced.

It is safe to say that there is a return on investment (ROI) when using project management. However, the return may not be readily apparent until two to five years after implementation. Like most activities, there are learning curve effects to be considered.

Solution Provider:

■ Project management processes allow companies to function as solution providers rather than merely as suppliers of products or services.

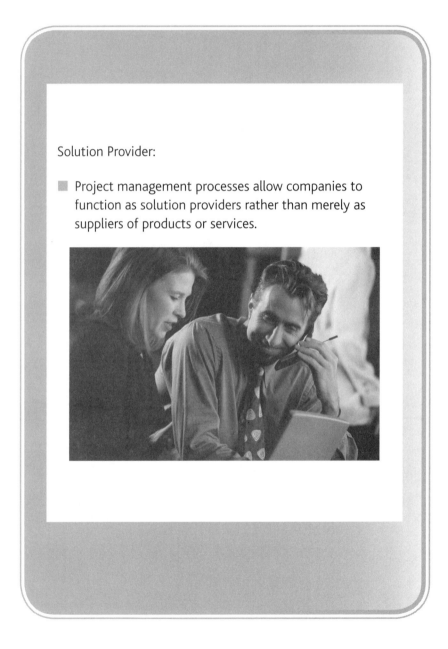

When companies excel at project management, they capitalize on this expertise by marketing themselves as business solution providers rather than as simply the seller of products or services. As solution providers, companies sell:

- Their project management skills

- The ability to interface with the company's project management methodology

- Their best practices library

In exchange for providing customers with business solutions, the company wants their customers to treat them as though they are a strategic partner rather than just a contractor. This concept of being a solution provider is now part of engagement sales and marketing activities. Solution providers focus on the long term rather than the short term; follow-on work rather than single sales; selling long-term value rather than specific features; selling quality of the solution rather than quality of the products; and giving customer service a high priority rather than a low priority.

Suboptimization Risks:

■ Project management allows us to make decisions that are in the best interest of both the company and the project.

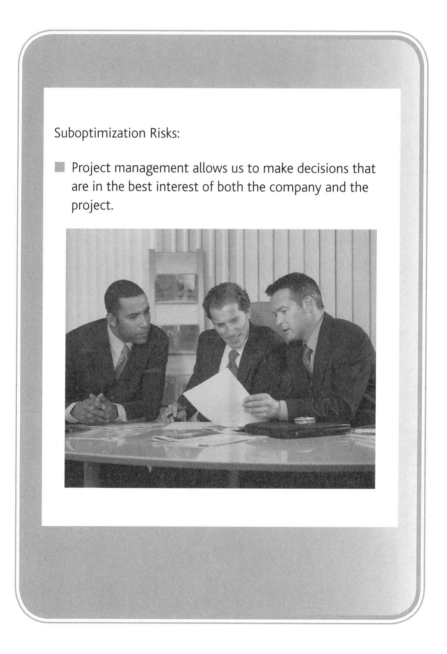

Individuals may have the tendency to make self-serving decisions, worrying about their own best interests rather than what is in the best interest of the company. Project teams must make decisions that are considered to be in the best interests of both the project and the company, and these decisions must be placed before their own best interests. However, this is not easily done. To achieve effective decision making, team members must understand the business as well as the technology. The less the business knowledge contained within the team, the greater the need for close interfacing among the executives, the project sponsor, and the project team.

It is the responsibility of the executive project sponsor to make sure that the team understands the business implications of the project. The sponsor also defines the business objective for the project and must support the project manager's efforts to ensure that the team makes appropriate decisions.

QUANTIFYING THE BENEFITS

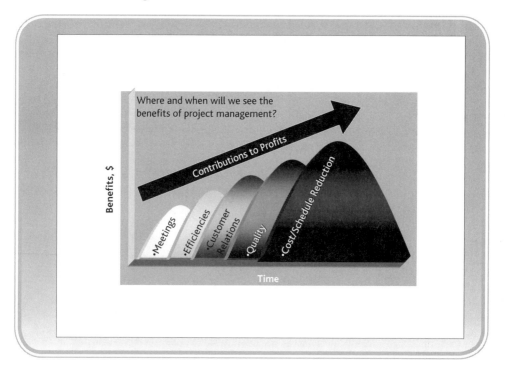

Selling executives on project management must address the quantification of the benefits and the timing of when the benefits will be seen. While the answer to these questions will vary from company to company, typical benefits may appear as:

- A reduction in the number of nonproductive meetings due to better teamwork and communication

- More efficient execution of projects due to the processes contained in the EPM methodology

- Getting closer to the customer, possibly resulting in sole-source contracting and a lowering of bidding costs

- An improvement in quality, resulting in greater customer satisfaction

The order and magnitude of these benefits will vary from company to company. Some companies engage in too many unproductive meetings and simply do not realize the cost implications. In such cases, reductions in the number of meetings can be highly cost effective.

Chapter

4

THREE CORE
BEST PRACTICES

THE FIRST BEST PRACTICE

Corporate recognition of professionalism in project management is necessary to maximize performance.

From any executive perspective, there are several best practices that should be considered. The first best practice is to create an environment where project management is viewed as a profession. People who enter the project management field want to be regarded as professional project managers and do not want to see project management viewed as a part-time profession. Part-time project management may create split loyalties and may not create the same high-quality results produced by full-time, dedicated project managers.

All too often, the recognition of project management as a profession is driven by the customers. When customers feel that project management certification is essential, contractors usually follow suit. Some companies today are mandating that all project managers identified as part of the competitive bidding process become certified in project management. This is why companies that are project driven or project based appear to have a greater need for certification than those companies that are non–project driven.

THE SECOND BEST PRACTICE

Given the importance of project management today and in the future, management must find ways to retain professionalism in project management.

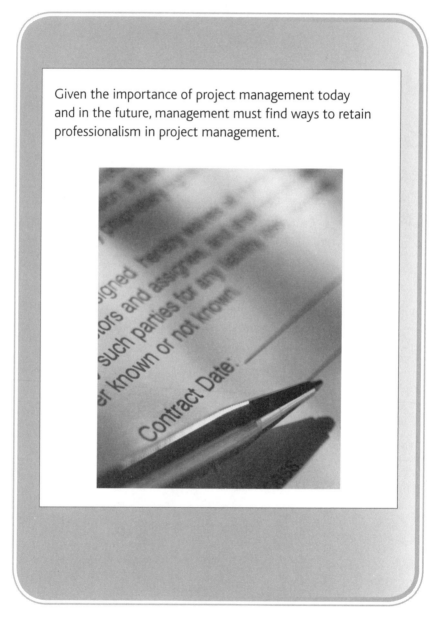

The second best practice is the retention of key personnel in pro-ject management positions. Not only do employees want to see project management as a recognized profession; they also want to see a project management career path within the company. Employees who want a career in project management will migrate to those companies where it is a recognized career path.

Making project management a career path may seem easy to do, but it comes with a critical problem: What specific job responsibilities should be placed in project management job descriptions to differentiate between pay grades or levels of project management? What information should appear in a project manager job description that will differentiate between grade 5, 6, 7, 8, 9, and 10 project managers? Using the dollar value of the project, project duration, risk levels, or experience with project management has proven to be ineffective.

There seems to be a tendency among human resource organizations to replace project management job descriptions with project management competency models. The competency models focus more on the skills that the individual should possess rather than the knowledge needed for a particular job.

THE THIRD BEST PRACTICE

Project management must be integrated and compatible with the reward systems for sustained project management growth to occur.

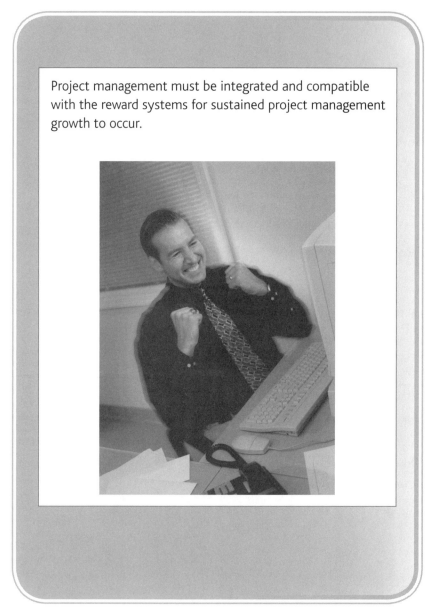

In the third best practice, employees want to believe that opportunities for rewards and advancement are possible through successful project management. It is an undesirable situation for employees to believe that the project manager position does not include promotion or other reward possibilities. If management does not offer the opportunity for people to be promoted as project managers, then:

- People will regard project management as just a part-time activity and rush to get back to their previous job, which may be a position that includes promotion opportunities.

- People will not be dedicated to the project, and the quality of the work may suffer.

- Customers who are external to the company may recognize this problem and may avoid agreeing to follow-on work.

- Capturing lessons learned and best practices in project management may become a low priority.

This does not imply that rewards should be forthcoming after the successful completion of each project. Rewards should be provided for continuous successful performance in project management. Without some sort of reward system, continuous improvements in project management may be difficult to achieve, thus limiting sustained project management growth.

Chapter

5

ROLE OF THE EXECUTIVE AS A PROJECT SPONSOR

HOW EXECUTIVES INTERFACE PROJECTS

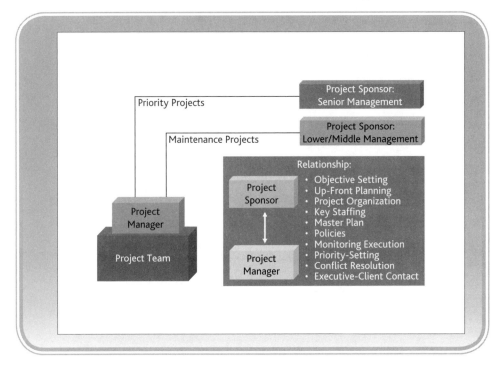

Previously, we stated that executives cannot divorce themselves from project management, but must be willing to function as an executive project sponsor. The illustration on the left shows the relationship between the project sponsor and the project manager. This illustration shows that:

- Sponsorship need not be at the executive levels all of the time.

- The sponsor exists for the entire project team rather than just the project manager.

- Many of the activities in the "Relationship" column can be the responsibility of the project manager, the sponsor, or both parties. However, the decision as to "who does what activity" can vary based on the life-cycle phase.

Sponsorship should exist for all projects. The role of the sponsor, other than to clear away roadblocks that appear in the path of the project manager, is to make sure that the business interests of the company are addressed. If an official sponsor is not appointed, then the project manager's immediate administrative supervisor may act as the sponsor by default.

THE EXECUTIVE SPONSOR'S ROLE

The executive sponsor maintains close contact with the project throughout the entire project life cycle. However, the actual responsibilities of the sponsor can change, depending on which life-cycle phase we are in.

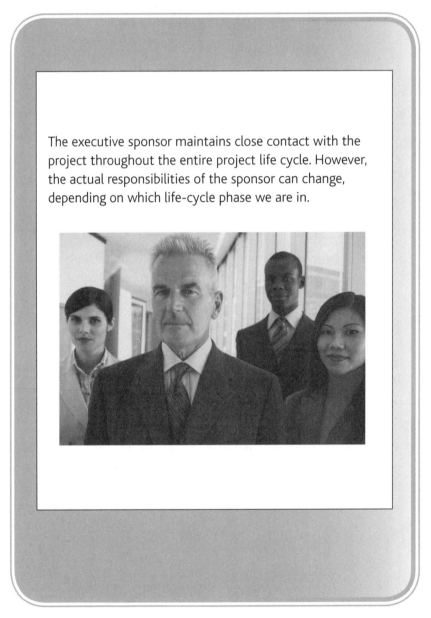

The role of the sponsor can change depending on which life-cycle phase the project is in. For example, during the project initiation and planning phases, the sponsor may take on a very active role to ensure that the proper objectives are established and that the project plan satisfies the needs of the business as well as the needs of a particular client. During the execution phase, the sponsor may take on a more passive role and get involved on an as-needed basis, such as when roadblocks appear, crises develop, and conflicts exist over priorities among projects.

There are other factors that can determine the involvement of the sponsor, including:

- The risks of the project
- The size, scope, and nature of the project
- Magnitude of the conflicts and obstacles
- Who the customer is
- Special needs and requests of the customer
- Quality and availability of assigned resources
- Priorities among projects
- Authorization for scope changes

For clients who are external to the company, major participation in the initial sales effort and contract negotiations must exist.

For projects being conducted for external clients, a contractual agreement must exist. Although project managers may possess some knowledge about contracting and procurement, the real expertise generally resides with the legal department and the executive project sponsors. Activities performed by the sponsor rather than the project manager include:

- Contractual negotiations

- The selection of subcontractors

- Identification of business- and non-business-related risks

- Assessing contractual risks

- Reviewing the terms and conditions of the contract

- The approval and signing of the contract

- Briefing the client's senior levels of management

- Selection of the project manager

- Determining the appropriate priority for the project

After contract award, preparation and endorsement of the project's charter occurs.

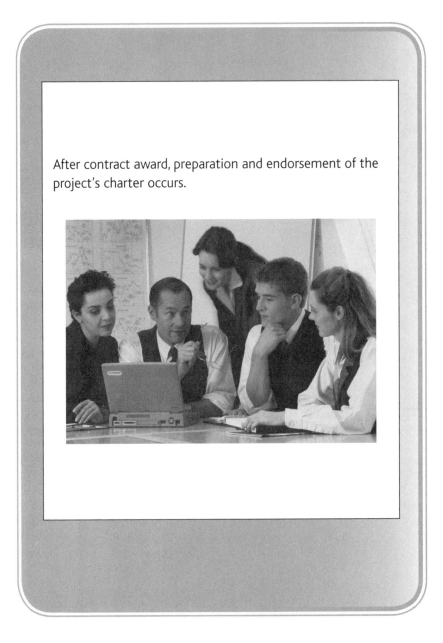

Once the contract is agreed upon, someone from the senior level of management is assigned as the project sponsor. The sponsor may have been identified in the proposal. The sponsor is generally responsible for the preparation of the charter, but this is often delegated to the project manager. However, the sponsor must sign the charter and endorse it before presenting it to the project team. Items usually included in the charter are:

- Project objectives and constraints

- Project assumptions and risks

- Identification of the project manager, including role, responsibility, and level of authority

- Project or product description

- Business need for the project

- Major players and key stakeholders

Every company has its own version of a charter. The information in the charter can vary based on the size and nature of the project, and whether the project is for an internal or external customer.

The Sponsor assists the project manager in getting the project under way (planning, procedures, key staffing, etc.).

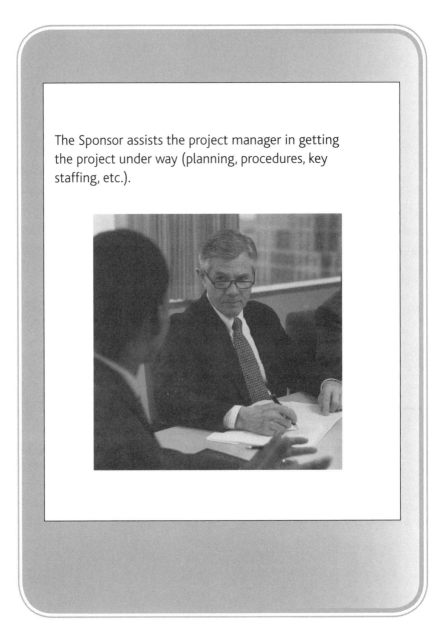

From a project management perspective, project initiation and planning are extremely critical life-cycle phases. Sponsors must make sure that these phases move along as smoothly as possible. Items that the sponsor provides support for include:

- Key staffing

- Interpretation of objectives, assumptions, and constraints

- Establishing the priority for the project and the reason for the priority

- Participating as needed in up-front planning

- Establishing policies and procedures unique to the project

- Assisting in the design of the organizational chart for the project team

- Introduction of the project team to the customer

Once again, the role of the sponsor and the accompanying formality can vary based on whether the project is for a customer that is internal or external to the company.

Making sure that the project has the appropriate priority and that both the project team and the functional (line) organizations know the priorities.

One of the items usually not included in the project charter is the priority of the project. The reason is that priorities can and do change over time. Priorities are established by executives and sponsors. This is accomplished before the project actually begins. The priority may have been established as part of the portfolio selection process.

The priority for the project has the single greatest impact on the selection of the quality and skill level of the functional employees to be assigned to the project team. While the project manager always wants the best possible resources, the established priority may dictate otherwise. When some projects are completed and other projects are initiated, priorities can change. And because of the changes in priorities, project managers must recognize that they may lose some of their critical resources.

Some project managers never negotiate for the best possible resources for fear of losing the resources if the priorities changes. The loss usually occurs at the most inopportune time. Some project managers would rather have average or above-average resources assigned, assuming they can do the job, and then retain the same people for the duration of the project. Continuity of personnel may be more important than fighting for the best resources.

Establishing the priority for the project (either individually or through other executives) and informing the project manager of the established priority and the *reason* for the priority.

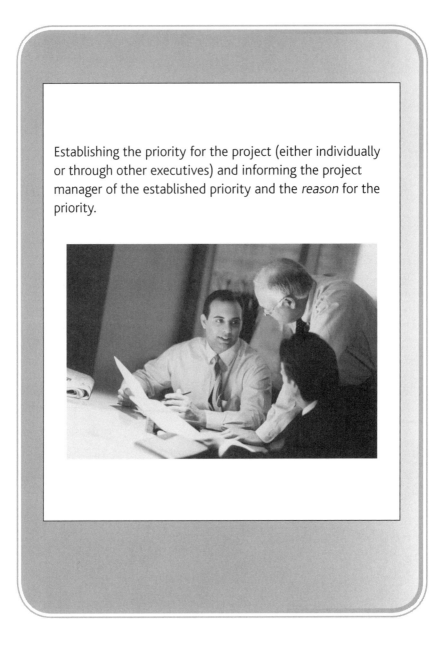

Establishing the priority for the project is and always will be an executive decision. Establishing the priority may be accomplished by the project sponsor, but it is most commonly established by the executives involved in the portfolio management of projects. If a project management office (PMO) is in place, then the PMO may also participate in providing recommendations for the priorities. It is then up to the project sponsor to disseminate the information.

Project teams are often disenchanted when they are informed of the priority level of the project, especially if it is a low priority. The project sponsor can solve this problem by personally informing the team about the priority level and the rationale behind the decision, at the same time discussing the importance of the project to the company. This is best accomplished during the initial kickoff meeting for the project.

High priorities may motivate teams more so than low priorities. However, high priorities can generate more stress for the team. Some companies that have hundreds of possible projects in the queue may prioritize them in groups of 20 at a time. In any case, all projects selected should be viewed as important to the organization.

Assisting the project manager in establishing the correct objectives for the project.

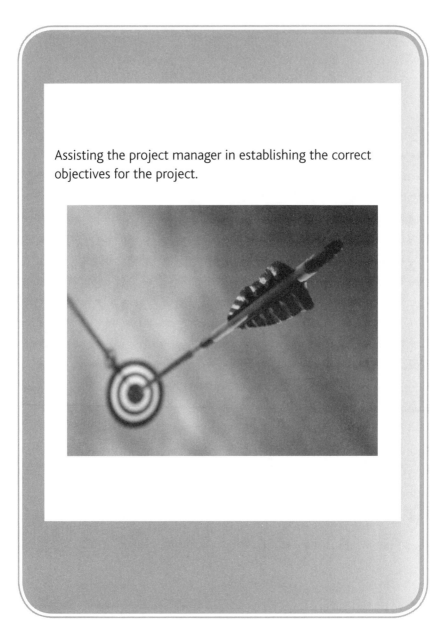

The objectives for the project are established prior to project initiation and in meetings between the customer, the project sponsor and, if assigned, the project manager. However, there may be hidden or alternative objectives established above and beyond the customer's objectives.

As an example, a company wins a contract to manufacture some products for a customer. On the surface, it may seem that the objective is simply the manufacturing of high-quality products for a customer. But now let's assume that, in order to win the contract, the company bid on and won the contract at a price 20 percent below its own cost of performing the work. The secondary objectives, not identified to the customer, might be to keep its people employed, look for future contracts from this customer, develop new manufacturing processes, or maintain a specific market share.

Not all secondary objectives are shared with the project team. Telling the team that the secondary objective was simply to keep people employed may create fear that layoffs are possible in the future. Sometimes secondary objectives are shared just between the project sponsor and the project manager.

Establishing and reinforcing the business objectives for the project as well as the technical objectives.

From the previous illustration, it should be obvious that there exist both technical and business objectives for a project. Technical objectives may emanate from the customer's statement of work, whereas the business objectives are established internally at the executive levels of management. Both the business objectives and technical objectives must be reinforced by the executive sponsor.

Some business objectives are company-proprietary objects and may not be shared with the project team. Although withholding information may be necessary, it does create problems when decisions must be made and the team has been given only partial information. In such a case, the team would focus on technical decision making and rely on the sponsor (and possibly project manager) for business-related decision making.

Making sure that all deadlines are realistic.

Project managers are usually not brought on board the project until after the contract is awarded. The result is that the sales personnel or contract negotiators may agree to project deadlines that are unrealistic. Project managers are then expected to perform miracles and accomplish the impossible. If the deadlines cannot be met, blame is usually placed on the project manager rather than the sales force that made the unrealistic promise.

The project sponsor must work with the project manager to make sure that all deadlines are realistic. Ideally, this should be accomplished before the contract is signed. Unfortunately, this is often not the case, and unwanted and risky trade-offs may be necessary.

There is a trend occurring in the project management community whereby project managers accompany sales personnel in initial meetings with the customer to make sure that what is promised can actually be delivered. This is highly beneficial but does require that the project manager be identified early on. Some companies are reluctant to do this because there is no guarantee that the contract will be awarded. The other issue is that even if the contract is awarded, the time frame of the contract may prevent this project manager from being assigned because of other commitments.

Reaffirming to the project manager and project team the importance of meeting deadlines.

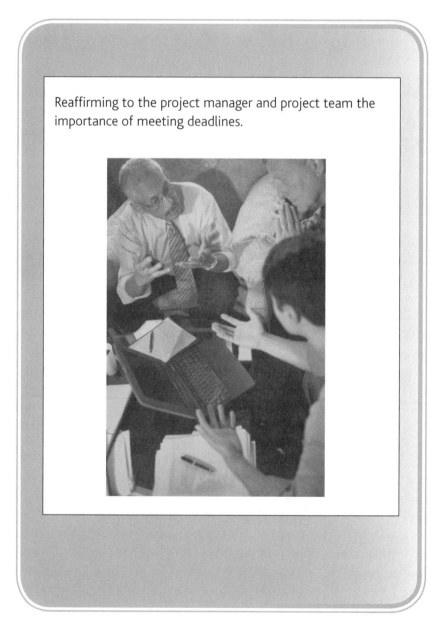

Assuming that the deadlines are realistic, the project sponsor must continuously reaffirm to the project manager and the project team the importance of meeting deadlines. Unfortunately, this is often difficult to accomplish because of changing priorities, the loss of critical resources, and unforeseen problems.

One possible solution to managing deadline issues is using the principles of risk management and the development of contingency plans. Team members must be encouraged to bring forth problems, especially issues with unrealistic deadlines, to ensure that the maximum number of options are identified for the development of effective contingency plans. Many companies encourage the swift identification and communication of project problems to ensure that corporate support can be forthcoming as quickly as possible. It is important to note that people must not be punished or criticized for bringing forth bad news.

There is an old saying in research and development (R&D): "If you can lay out a project plan for R&D, then you do not have R&D." In some projects, such as with R&D, it may be impossible to commit to a deadline at project initiation because of the many unknowns. Deadline commitments may have to be periodically reevaluated, and trade-offs may be necessary to meet the initial agreed-upon deadlines.

Explaining to the project manager the environmental/ political factors that could influence the project's execution.

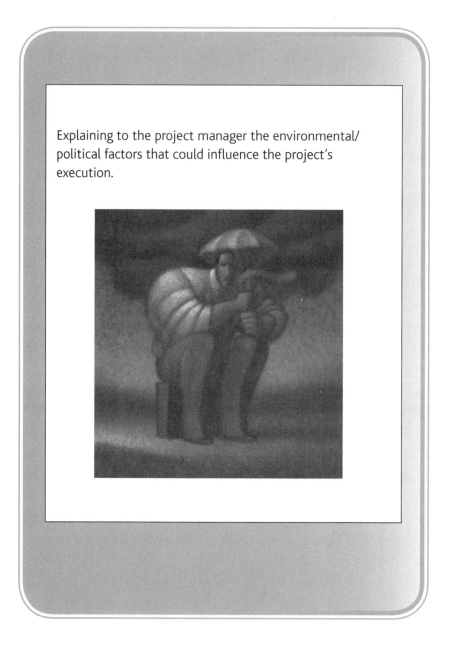

The *Project Management Body of Knowledge*® (PMBOK) *Guide* refers to a planning input known as *enterprise environmental factors*. Simply stated, these are factors that are internal and external to the organization that can affect the accomplishment of project objectives, the way in which the project is planned, and the project decision-making process. These factors, which include organizational culture and resource capabilities, can also change over the duration of the project.

One of the roles of the executive sponsor is to assist the project manager in the identification of these factors and how they may affect the project. Not all project managers have a strong business knowledge or are considered to be business savvy, and many newly appointed project managers may not have a grasp of the many issues and complexities of their own industry. Guidance by the sponsor is essential.

Some of the guidance that sponsors provide may include:

- Inflationary impact on corporate spending

- Future products or services identified in the firm's strategic plan

- The firm's hiring and layoff intentions

- Potential merger or acquisition opportunities

- Selection of preferred or strategic suppliers and partners

- Upcoming negotiations with labor unions

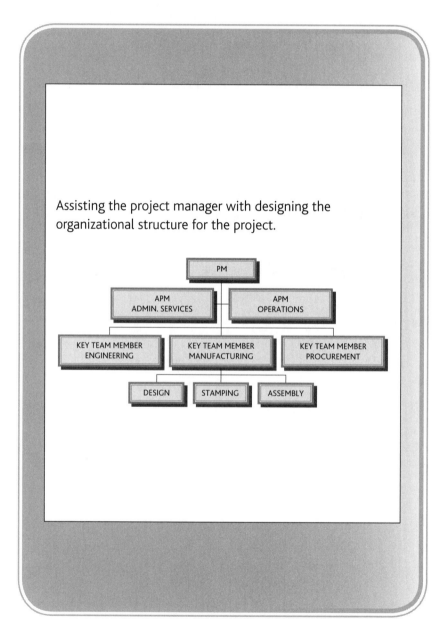

Assisting the project manager with designing the organizational structure for the project.

The executive sponsor will assist the project manager with the designing of the organizational structure for the project team. Factors that must be considered include:

- Amount of interfacing with the customer
- Skill levels needed
- Availability of personnel
- Ability to retain the people for the duration of the project
- Full-time versus part-time assignments
- Funding available from the customer
- Number of customer interface meetings
- Timing and content of project reports
- Proximity to the customer
- The customer's involvement in day-to-day project activities
- Technical complexity of the project
- Amount of functional integration required

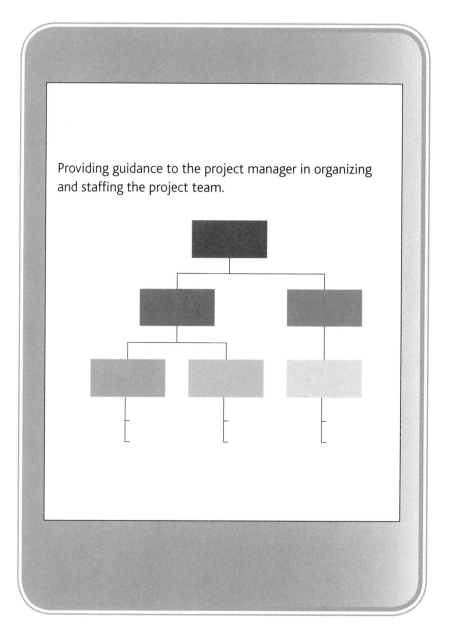

Providing guidance to the project manager in organizing and staffing the project team.

Involvement by the executive sponsor during organizational team design and staffing can make the management of the project much easier for the project manager after execution begins. The executive sponsor may have knowledge related to:

- What type of project structures worked well on previous projects and what structures worked poorly?

- What project team structures worked well for this client on previous projects?

- Is there sufficient funding set aside for this type of team structure?

- Will this team structure set a good or poor example for the structure of future teams?

- Will the length of the project impact the design of the structure?

- Will the risks in the project impact the design of the structure?

- Will the customer be permitted to have a say in the design of the structure?

- Will the structural design be impacted by the number of and importance of the stakeholders?

Assisting the project manager with filling the key staff positions for the project, both at project onset and in an emergency.

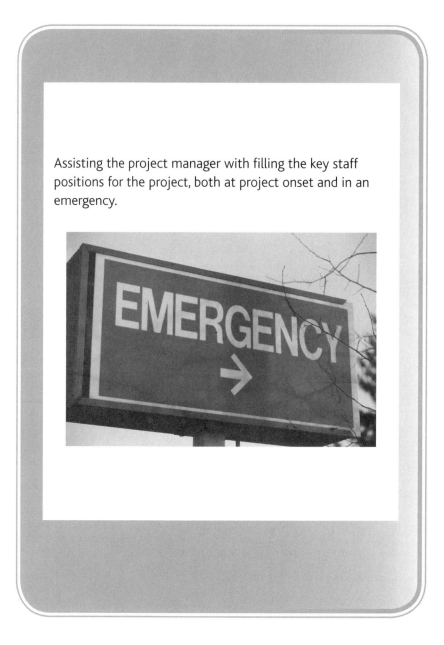

More and more companies today are using governance and portfolio management techniques for selecting projects, determining priorities, and deciding in what sequence projects will be implemented. Many times, the portfolio selection process is based on capacity-planning activities, where the projects selected are matched against the availability of specialized subject matter experts.

Project managers may have no knowledge about capacity-planning activities and the project selection process. However, the executive sponsors generally have this knowledge and can assist the project managers in staffing some of the key positions. Some of the critical decisions that the sponsor can assist with include:

- What skill levels are necessary?

- Will the resources be needed full time or part time?

- Will there be a risk of losing key resources to other projects?

- Is it better to have the same people assigned for the duration of the project or to negotiate for the best people available?

- What is the authority level of the project manager regarding how functional resources will be managed?

- Will the resources be working on multiple projects, and how will priorities be managed?

Helping resolve conflicts between the project manager and line organizations.

Regardless of the decisions made during the portfolio management of selected projects and the establishment of priorities among projects, line managers may still find it difficult or even impossible to assign appropriate resources to projects. Possible reasons might be:

- Resources are still required on existing projects

- Critical resources have left the company

- Resources must be used to solve crises or put out fires on other projects

- Changing priorities may cause a shift in resource assignments

Sponsors must step in occasionally and help resolve conflicts between project and line managers. In this regard, sponsors may function as referees. Not all projects can receive the best functional resources available. Sometimes it is better for the project managers to negotiate for deliverables than to negotiate for specific resources. Project managers may request specific resources, especially if they have previous working relations with these people, but there is no guarantee that the line managers can or will assign these people.

The success of project management is heavily based on a collaborative working relationship between project and line managers. Conflicts should be brought up to the sponsor quickly for resolution.

Interpreting company policy for the project manager.

It is almost impossible to design company policies such that they cover all possible situations that can exist on each and every project. In addition to these issues, company policies are subject to interpretation and misinterpretation.

The role of the sponsor is to assist the project manager and team with this interpretation. The sponsor also serves as a "safety net" for the project manager should any of the policies require a specific interpretation for the benefit of the project. Project managers may also be required to adhere to some of the client's policies and procedures, and the sponsor can assist in this interpretation as well.

Most companies today are in a paperwork reduction mode and focus heavily on forms, guidelines, templates, and checklists. These documents may also require interpretation, and the sponsor can be of assistance here as well.

Providing guidance for the establishment of policies and procedures by which to govern the project.

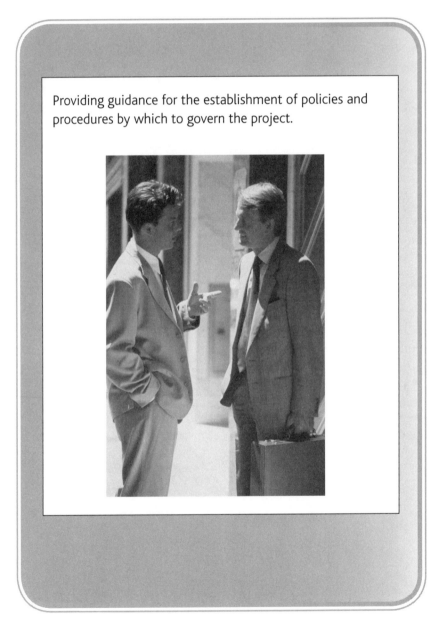

It is possible for long-term or multiyear projects to require the development of specific policies and procedures by which the project will be governed. Reasons for having project-specific policies and procedures include:

- Special requirements imposed by the client

- Risks of the project

- Control of proprietary or confidential intellectual property

- Specific interfacing requirements with the customer

- Specific interfacing requirements with the suppliers

- Reporting of test results, including bad news as well as good news

- Timing and details of project reporting

- Disclosing information to the media

- Disclosing information the some of the client's competitors

Project sponsors must work closely with the project manager to establish these project-specific policies and procedures.

Helping the project manager and project team establish project-specific policies and procedures that are acceptable to both the parent company and the project team.

Project managers would like nothing better than to establish their own policies and procedures for long-term projects. However, there may be risks that project-specific policies and procedures may be in conflict with the parent company's established policies and procedures.

Project sponsors must make sure that the compatible policies and procedures are in place. When significant differences occur, the sponsor will work to negotiate solutions and once again provide a "safety net" or possibly some degree of protection for the project manager and project team.

Allowing project managers to develop their own project policies and procedures can lead to significant conflicts and unfavorable results, especially if all of the project managers opt to do this. The result could be the opening of a "Pandora's Box" of project issues. Sponsors must make sure that there is a genuine need for the development of customized policies and that this practice is kept to a minimum.

Assists project manager in identifying and solving major problems.

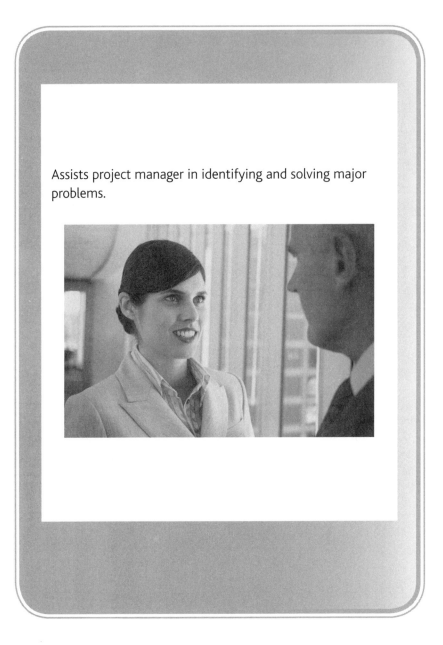

As mentioned previously, project managers generally possess limited authority, and this can affect their ability to resolve problems that are encountered during project planning and implementation. Virtually all problems that are experienced are directly or indirectly related to the triple constraint of time, cost, and performance.

Sponsors have the authority and ability to make immediate decisions. Sponsors also have direct access to the executive levels of management, thus accelerating the decision-making process. Problems the project managers cannot resolve without executive support include:

- Increased funding for the cost baseline

- Major issues that may negatively impact customer satisfaction

- Sole-source subcontractor selection

- Increasing the size of the project team

- Changes to the project's priority

- Removing functional employees from the project team

- Rewards provided to the project team

- Managing policies regarding the use of overtime

Helping the project manager put out fires or resolve conflicts in a timely manner.

Decision making in most companies is a tedious process. Because project managers may have limited authority, they may be at the mercy of others when it comes to decision making. Also, project managers may not be familiar with all of the factors that must be considered in the decision-making process.

Employees and executives alike may not realize the importance of the time constraint on a project and the effect of the decision making process. The sponsor has the authority to accelerate the process to minimize impact to the project. On one government contract, a "Unit Manning" document existed, specifying the number of employees and skill levels that could be assigned to the project. The project manager was in desperate need to have an additional person assigned to the project. It would have taken the project manager six months to have the Unit Manning document changed. The sponsor worked directly with the human resources executive and accomplished the change in two weeks.

Establishing and maintaining the executive-to-executive client relationships with both internal and external clients.

In the early years of project management, military officers, such as colonels, believed that the project managers in the contractors' organizations were beneath them in rank and therefore channeled all questions and communication to the executive levels of management. In some cases, executives were forced to interface directly with external customers. Customers would call every executive available until they received an answer that was acceptable.

To alleviate this problem, and to make sure that there was only one official company position regarding the project, the project sponsor position was created. It was common practice during competitive bidding activities to include the resume of the project sponsor in the proposal and to clearly delineate to the customer that all executive-level communications were to go through the sponsor or designated individual.

Another benefit can be found in the quality of the information that would be exchanged. Project managers often address issues involving minor details that the customer's executives are not interested in. Sponsors generally have a better understanding about the information that their counterparts wish to receive.

Functioning as the contact point for other executives in the parent company.

Project managers do not necessarily brief each and every executive in their own company about the status of the project. Instead, the project manager briefs the project sponsor associated with the project, who in turn briefs the other executives. One of the reasons for doing it this way is that project managers often get bogged down in the details that may not be of significant interest to the executives. Project sponsors generally have a good grasp of the information that other executives wish to hear.

Assisting the project team with the up-front planning for the project, especially the project management plan.

As mentioned previously, project sponsors are expected to be actively involved in the early phases of the project in order to verify and assure that:

- The project objectives are clearly defined and understood.

- The business reason for undertaking the project is communicated and understood.

- The priority for the project is explained to the team.

- Up-front planning is accomplished correctly.

- A master plan for the project is established and satisfies the customer's milestones.

Up-front planning focuses more on the project management plan than the detailed project plan. Because the team may be composed of individuals who have never worked together previously, questions will naturally appear over roles, responsibilities, customer interfacing, reporting, and other such issues. Although the project manager may be capable of handling these questions and issues, the presence of the project sponsor can facilitate the communications process, save time, and resolve conflicts quickly.

Assist the project team in developing the master plan for the project.

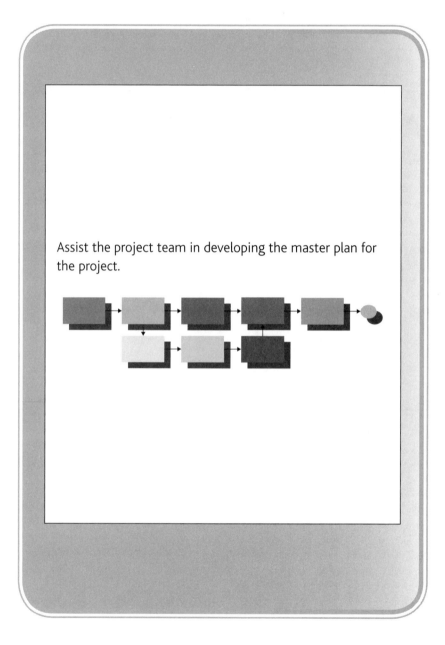

Once the project management plan is developed, the next step is the preparation of the project's master plan. This is a high-level plan including critical milestones established both internally and by the customer.

The sponsor's involvement is basically on an as-needed basis. The sponsor also validates the high-level plan to make sure that all critical milestones are identified. The master plan identifies the critical milestones for the detailed planning process and is usually presented as a milestone chart, but can also appear in the form of a high-level network diagram with time-phased milestones superimposed on it.

The master plan will, in most cases, require the approval of the customer to make sure all critical milestones are identified. The presentation of the master plan to the customer may be made by the sponsor as part of executive-to-executive communications.

Clearly delineating what expectations they have of the project manager.

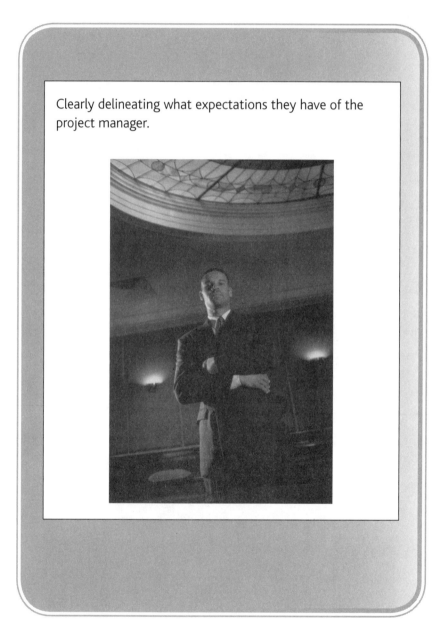

Every sponsor has his or her own management style and ideas about what an effective project manager–project sponsor relationship should be. The project sponsor must clearly delineate these expectations to the project manager. This could include frequency of briefings and the process to be used when problems and critical issues should be elevated to the senior levels.

Likewise, the project manager may wish to identify the expectations that he or she has of the sponsor. This could include availability for meetings and support for conflict resolution. In later sections, we will go into more depth on these expectations.

Clearly delineating what skills project managers are expected to possess and what they may be expected to do for lifelong learning.

Project sponsors may serve as mentors for the project managers. As such, the sponsor should identify what leadership skills are expected. If deficiencies are identified, the sponsor might provide additional coaching and recommend personal development programs or a lifelong learning program for project manager growth.

The lifelong learning process may include:

- Encouragement to attend seminars on general or specific project-related knowledge

- Encouragement for advanced degree coursework in project management

- Encouragement for advanced degrees in business, if appropriate

- Encouragement of membership in professional organizations

- Volunteer work as a company representative

- Encouragement to become a Project Management Professional® (PMP) or other project management credential

Providing the project manager and project team with directions on how to handle proprietary information.

Project sponsors must provide guidance to the entire project team about how to handle proprietary knowledge and intellectual property. This includes:

- How to identify proprietary information

- How to decide what proprietary information, if any, should be presented to the customer

- How to store and safeguard proprietary knowledge

- How to dispose of proprietary knowledge when necessary

These items focus on internally generated proprietary information. The contract may require that the customer provide the contractor with proprietary information, and this data should be handled with the same care.

Government contracts may require that the contractor abide by government policies and procedures with regard to intellectual knowledge and security restrictions. This may also require scheduled or unscheduled audits by the government to determine how security is being maintained. Some companies have an internal security department to support all project teams.

Make sure that there exists a structured process for the authorization of customer requests after go-ahead.

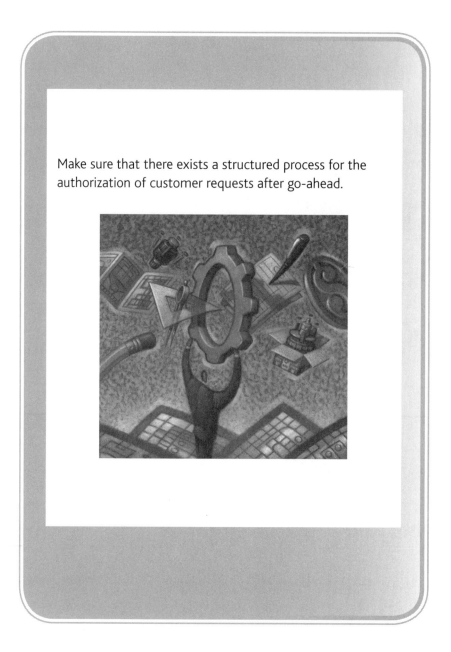

Both project managers and project team members often try to appease the customer by agreeing to no-cost scope changes. No-cost scope changes do not exist. Resources are required for any scope change, and funding must be made available for resources involved in the changes.

The project sponsor must encourage and enforce the use of a scope change control process. The critical questions to be addressed include:

- Cost of the change

- Impact of the change on the schedule

- Anticipated added value for the customer

- Risks associated with the change

Project sponsors must approve or at least be notified about all out-of-scope work. The approval of the scope change may result in a modification of the scope baseline. In any event, documentation must exist and be archived to provide an audit trail for all scope changes to the original contract.

Authorizing and approving contractual changes after go-ahead.

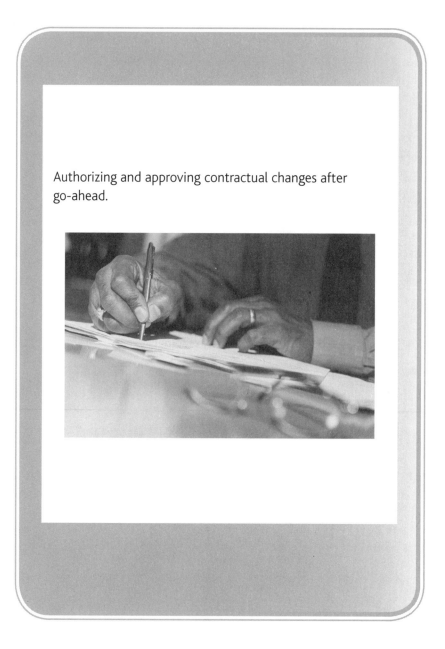

While project managers can identify the need for possible scope changes, they may not necessarily possess the authority to approve and sign off on the change. Changes may be amendments to the original sales contract or possibly even a new sales contract. Signature authority generally resides with senior management. Officers of the company or possibly the legal staff may be the only people authorized to approve changes.

The process by which changes are approved is established early in the project life cycle and managed specifically by the change control board, which is usually composed of key stakeholders that have a major interest in the project and have voting rights. In situations where a change control board exists, final signature authority may rest with the project sponsor.

Monitoring the high-level progress throughout the life cycle of the project.

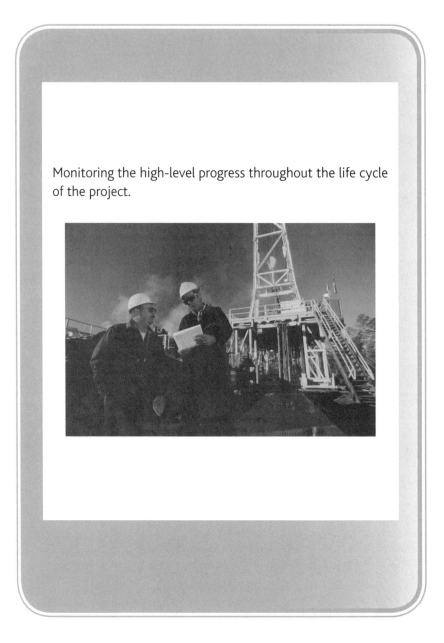

Acting as a project sponsor is just one of the responsibilities of senior management. Sponsors may actually have line function responsibility that consumes most of their time. Therefore, sponsors do not have sufficient time to become actively involved in the details of the project especially after project execution begins.

Sponsors are expected to monitor project progress from a high level of the work breakdown structure (WBS) rather than the lower, more detailed and technical levels. Project managers may meet with the sponsor on a regular basis to provide high-level briefings. While it may seem appropriate for sponsors to attend team meetings, this may become an issue regarding the effective use of the sponsor's time, especially if the executive is sponsoring multiple projects concurrently.

Some companies are using traffic light status reporting for projects to simplify information distribution:

- *Green light.* Work is progressing as planned; sponsor involvement is not necessary.

- *Orange or yellow light.* Work is progressing as planned, but a potential problem may exist.

- *Red light.* A problem exists, and involvement by the sponsor may be essential.

Traffic light reporting can minimize the number of meetings with the sponsor.

Sometimes project managers are young and/or inexperienced. The project sponsor must have patience in dealing with them.

An executive once commented to a newly appointed project man-ager, "The first time you make a mistake, I, the sponsor, will assume responsibility. If you ever repeat the mistake, you may get fired."

This statement illustrates that mistakes will happen, especially with young or inexperienced project managers. Executives must have patience in dealing with them. Establishing a mentorship program may be one way to minimize these mistakes. Another way is through training and education programs where project managers can learn from the mistakes of others rather than through their own mistakes.

Because of the demands made on the sponsor's time, mentorship may fall under the responsibility of the PMO, assuming one exists. It is often easier for the PMO to perform the mentorship as long as the issues relate to project management. Issues outside of the realm of project management may require the mentorship of the sponsor.

Sponsors must make sure that project managers understand that the role of the sponsor does not include day-to-day involvement with the details of the project. The exception is during a crisis.

All too often, project managers do not fully understand the role of the sponsor. Some people believe that the sponsor should be used as a dumping ground for all issues that the project manager does not want to handle. Others believe that sponsors should be involved in all of the day-to-day details of a project.

After project initiation, project sponsorship is more passive than active. Project managers are expected to use good judgment when bringing problems and issues to the sponsor for advice and guidance, but ownership of the problem remains with the project manager, not the sponsor. Project managers who need continuous supervision and guidance from the sponsor or executive are often replaced with people who have the ability to manage more effectively.

While some project sponsors can and often do participate in solving problems, their primary role is to manage from a higher level and provide information and advice to the project manager. Sponsors do not have the time to solve all problems for the project managers. There is an old adage that relates to this situation: "If you go to the well too often, eventually the well will dry up."

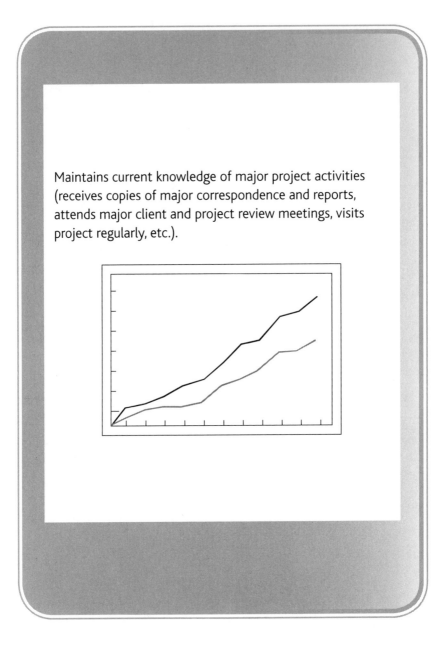

Maintains current knowledge of major project activities (receives copies of major correspondence and reports, attends major client and project review meetings, visits project regularly, etc.).

After project initiation and planning is completed, the role of the sponsor becomes more associated with monitoring and (some) control and, of course, conflict resolution. Ongoing activities for the project sponsor include:

- Being briefed on major project activities, especially those that are necessary for executive-to-executive contact with the client

- Receiving copies of major correspondence and reports from both the project manager and the client

- Possibly reviewing all formal letters sent to the customer that may relate to proprietary knowledge, scope change, and other contractual changes

- Attending critical project review meetings

- Attending end-of-phase review meetings, especially if a go or no-go decision must be made

- Occasionally visiting the project team to provide encouragement and support

Sponsors (and possibly the customer) should plan on attending project management training programs so that they and the project team are speaking the same language.

The role of the sponsor should not be left to chance. Many companies have developed half-day training programs to provide executives with information and suggestions about how to fulfill the project sponsor role. The secondary benefit of these training programs is to provide sponsors with knowledge of common project management terminology, such as work breakdown structures, statements of work, and network diagrams.

Some companies invite their customers to attend these training programs. The primary reason is for the customer to clearly understand the role of the sponsor. Today, many executive sponsors have carried this one step further by becoming certified as a PMP®.

Companies that survive on competitive bidding often insert the resume and job description of the project sponsor into their proposal. There are two major reasons for this:

- To advise the customer that someone from the senior levels of management will be overseeing the project and how the project manager manages costs and schedule

- To make the client familiar with the sponsor's role and to establish the sponsor as the executive-to-executive contact point

Sponsors must understand that effective project management is more than just a schedule or a budget.

All too often, executives focus too heavily on the triple constraint and believe that project success can be measured by analyzing schedule and budget performance. While there may be some merit in this, there are other factors that may not be important to the project manager but should be monitored by the sponsor:

- Making sure that the work flow of the project does not disrupt the ongoing work in the company

- Making sure that the project's work flow does not alter the corporate culture

- Adherence to ethical conduct standards

- Maintaining compliance with government regulations for health, safety, and environment projection

- Making sure that proprietary knowledge needed in the future is documented and retained

- Making sure that the customer's definition of *success* is the same as the project manager's definition of *success*

Sponsors should provide constructive feedback to the project manager and project team in such a manner that it is not viewed as personal criticism.

Mistakes will happen during project planning and execution. Possible reasons for the mistakes include:

- Misinterpretation of requirements or instructions

- Inexperienced workers

- Failure to assess risks or agreeing to a high-risk shortcut

- Excessive and/or prolonged overtime

Sponsors and project managers are expected to provide constructive feedback. However, the feedback should be provided in such a manner that employees recognize that it is constructive feedback and not personal criticism.

Both project managers and project team members desire feedback. People want to know what they are doing well and what needs to be improved or done differently. Providing this feedback in a non-constructive manner can alienate team members and may have a serious negative impact on morale.

Keeps general management, company management, and possibly the media advised of major problems.

For some industries, the larger the project, the greater the attention provided by the news media. It has been said that bad news sells more newspapers than good news. Schedule delays and cost overruns on highly visible projects may make the front page.

Project sponsors are generally more qualified or in a better position than project managers to address the media in times of crisis. Project managers view crises from a project perspective, whereas sponsors look at the same issues from a company perspective. Also, customers expect to hear about critical problems first from the sponsor rather than the project manager. The reason for this is that it provides the customer with some degree of satisfaction that senior management is involved and on top of the issue.

Encourage the project manager and team members to bring forth bad news as well as good news.

Very few projects are completed without experiencing some bad news. The real issue is the timing of when the bad news is brought to the surface. Some sponsors set an expectation that:

- They want to hear only good news

- They do not want to hear bad news unless it is validated

- They do not want to hear bad news without also hearing the corrective action plan

These conditions actually can generate greater problems. Executive sponsors should encourage project managers to bring forth bad news (and good news) as soon as it is known. The quicker the news is known, the faster options can be determined and become available for corrective action. Also, sponsors may know of options that are unfamiliar to the project managers.

Executive sponsors must not "lose their cool" when bad news appears.

Sponsors must be willing to accept the fact that bad news will happen. When bad news appears, the worst thing sponsors can do is to "lose their cool" and overreact. If the sponsors demonstrate a trend of overreacting to bad news, then the project managers:

- Will hide bad news for as long as possible

- Will not bring bad news forth in a timely manner

- Will bring the bad news forth at a point where very few options exist for corrective action

- Will try to solve the problem themselves, possibly by using the worst alternative or least cost-effective solution

- May take unnecessary risks to resolve the problem and expose the company to additional problems

- May run the risk of having the customer hear about the bad news before the sponsor hears about it

- May have the news media hear about the bad news before either the sponsor or the customer knows of the problem

Given that bad news is inevitable, there are ways to handle it by which team morale will not be damaged.

When sponsors overreact to bad news, people tend to bury problems rather than bring them to the surface for resolution. Sponsors must work with project managers to develop a project policy or guideline for how problems and issues will be handled. This must be done in a manner such that morale will not be damaged. Possible approaches include:

- Encouraging people to bring bad news forth as quickly as possible
- Eliminating the perception that people will be punished for bringing forth bad news
- Encouraging people to develop and communicate alternatives and recommendations along with problem situations
- Establishing a crisis team or crisis committee for evaluating bad news and project issues

The process for communicating and managing bad news should be included in the project management plan.

One of the most dangerous situations occurs when the sponsor believes that he or she is also the project manager.

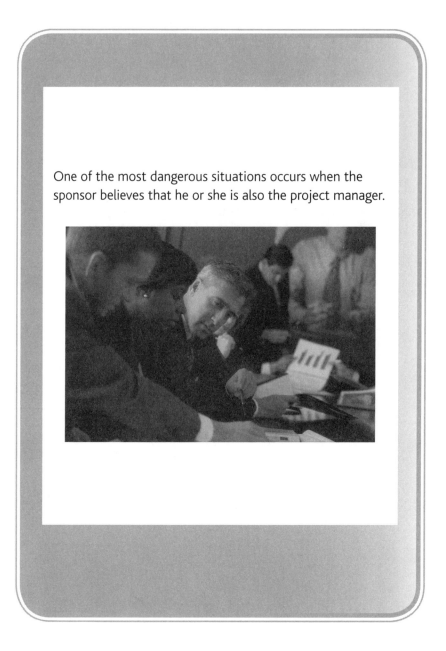

One of the most dangerous situations occurs when the project sponsor believes that he or she is the actual project manager, making all project decisions, and the real project manager simply executes all of the decisions made by the sponsor. When this type of micromanagement occurs, the result is usually:

- Suboptimal decision making.

- Poor team morale.

- A desire by team members to leave the project.

- Lack of motivation.

- Problems are not communicated or are ignored.

- No one accepts responsibility and/or takes a leadership role when a problem occurs.

- All issues are brought to the sponsor regardless of the magnitude of the problem.

- Team members refuse to make decisions, no matter how minor or mundane.

Sponsors must recognize the telltale signs that there may be excessive pressure placed on the project team.

By definition, projects are often unique endeavors, the type of which may not have been accomplished previously. As such, project managers and team members may not have any reliable information or basis for assessing the workloads to be placed on them in the fulfillment of the project objectives or contract. The result could be a much greater workload and effort required to complete objectives. This could be of great significance since it is well known that excessive workloads can lead to stress, burnout, and numerous mistakes.

Project sponsors must be able to recognize the telltale signs that the project team is under excessive pressure and stress. Team members should be encouraged to identify their own energy cycle and limitations and balance the workload effectively. A common practice is to perform the most difficult activities during periods of peak energy. Project managers should also identify the team's energy cycle to determine when and how work will be assigned and when it is best to hold team meetings to maximize effective decision making, team building, and obtaining support for decisions.

Sponsors must make sure that project managers understand that project managers are both managers and doers.

Inexperienced project managers often perceive the role of the project manager as sitting behind a desk and waiting to make important decisions. In reality, project managers are both managers and doers. Unless the project manager has an abundance of resources, he or she may be called upon to perform some of the work. The difficult part rests in deciding what work the project manager can actually perform while managing the higher-level integration issues associated with the project. Project managers must maintain an awareness that other team members may be able to perform certain work activities faster and better, assuming they have the available time.

Sponsors must understand the cost of paperwork or its electronic equivalent.

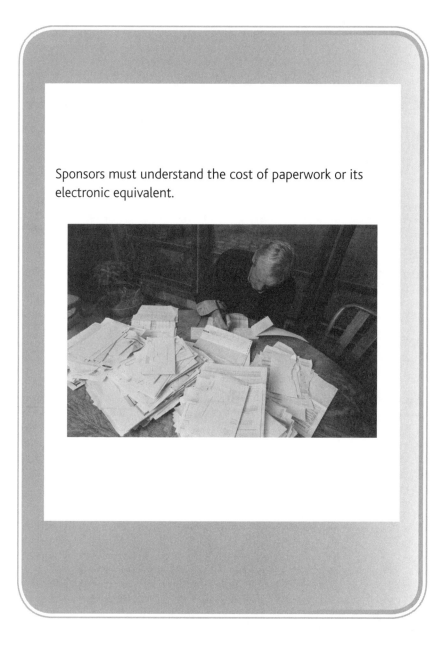

Perhaps the greatest pitfall that sponsors may encounter is the lack of knowledge about the cost of paperwork or its electronic equivalent. There is no such deliverable as a no-cost report. The steps included in preparing a report include organizing, writing, typing, editing, proofing, retyping, preparing graphic arts, approvals, sign-offs, reproduction, classification, distribution, storage, and disposal. In other words, there is time and effort involved.

It is not surprising that 8 to 10 hours of labor may be needed to perform all of the steps required for each page of the report. At $120 per fully burdened hour, the cost per page could easily reach $1,200 per page. Paperwork cannot be eliminated but should be kept to minimum levels.

The ultimate goal might be paperless project management with a reliance on automatically generated reports. This can be partially accomplished using the traffic light reporting system discussed previously.

Sponsors must understand that not only does someone have to write the reports, but someone must also read the reports.

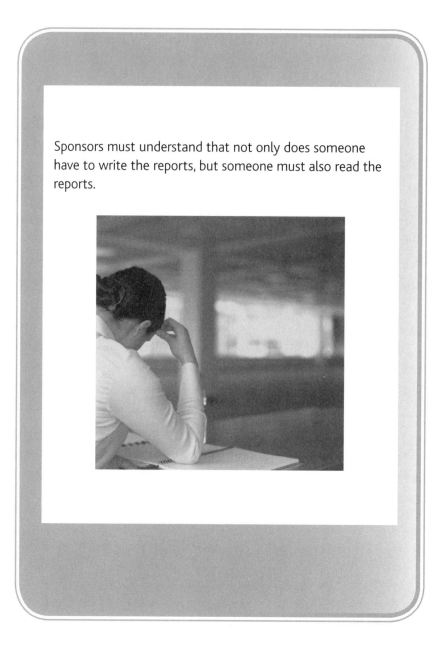

In the previous example, we discussed the cost of creating paper-work. Another cost that is often overlooked is the cost of reading the reports. Someone must read and possibly respond to the report. If this is a report for the senior levels of management, a fully burdened hour will be significantly more costly than $120.

While some type of paperwork is required for most projects, too much paperwork removes employees from other, more essential activities. A common rule or belief within the business environment is that employees spend six hours per day doing actual work and two hours per day involved in paperwork. When companies try to reduce the paperwork requirements, the productivity levels usually increase and can approach seven or more hours per day.

The only real way of determining the true status of the project is with walk-the-halls management.

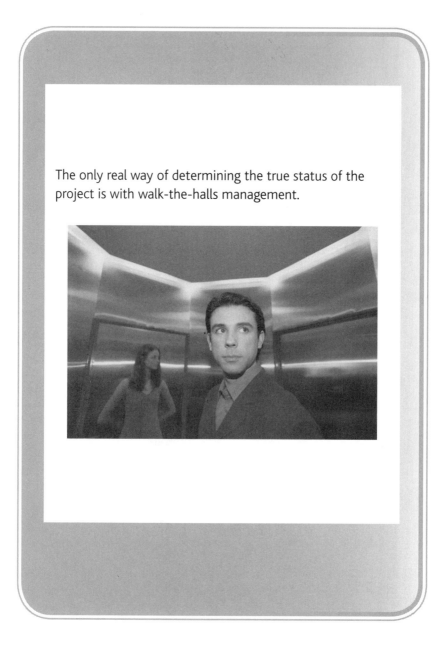

Reports, no matter how detailed, do not always show the true status of the project. Real status can be seen using the walk-the-halls or management-by-walking-around concept. In addition to seeing the true status, another benefit experienced is the motivational impact on team members when they see and interface with sponsors who display a genuine interest in the project and the project team.

Team members want to believe that that the project manager and sponsor are sincerely interested in the work they do. Not all team members attend project team meetings and, as a result, may work on a project for an extended period of time and never actually see or speak with the project manager or sponsor. This can have a negative effect on personal and project performance.

Sponsors must understand that the more meetings the team must attend, the less time they spend at their desk.

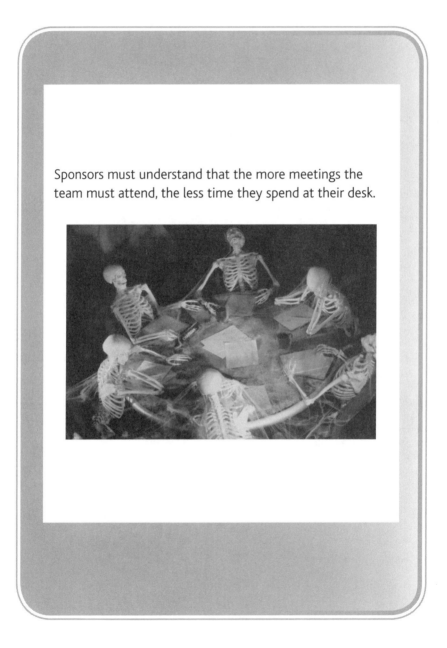

In the early years of project management, there existed a mistaken belief that the more life-cycle phases that existed in a project management methodology, the better project management would perform. It did not take long to discover that project managers were spending more time preparing for the life-cycle gate review meetings than actually managing the project.

While some meetings are essential, all meetings should be managed effectively, and meetings in general should be kept to a minimum. There should also be a process in place for evaluating the effectiveness of meetings by asking the following questions:

- Was the meeting necessary?

- Were the right people in attendance?

- Was the meeting productive?

- Could the meeting have been done using a medium other than personal attendance?

- What was accomplished?

- How can future meetings be improved?

Sponsors must understand that there are many forms of bureaucracy, which results in significant red tape for the project team.

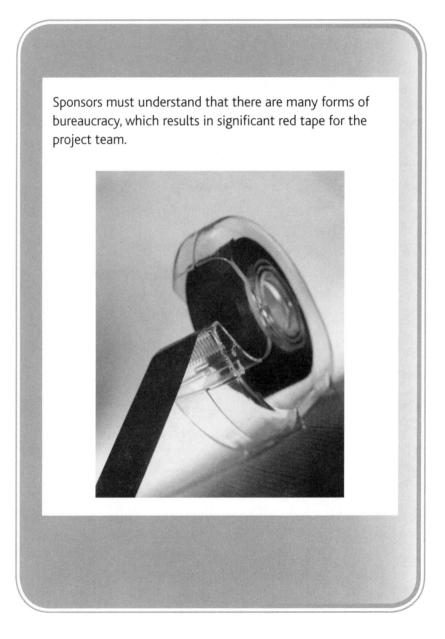

Previously, we identified two forms of bureaucracy: too many meetings and too much paperwork that may prevent project managers from performing more productive work. Project sponsors must try to reduce project red tape and clear the way for greater productivity. Possible activities that sponsors can perform to reduce red tape include:

- Limiting the number of levels of review

- Limiting the shifting of priorities

- Avoiding procrastination and the layers associated with decision making

- Eliminating or correcting unclear roles and responsibilities

- Educating customers and other stakeholders

- Eliminating vague goals and objectives

- Preventing department "buck passing"

- Eliminating a desire for perfection

- Protecting the team from company politics

Sponsors must ease the tug-of-war battles between the customer and the project team.

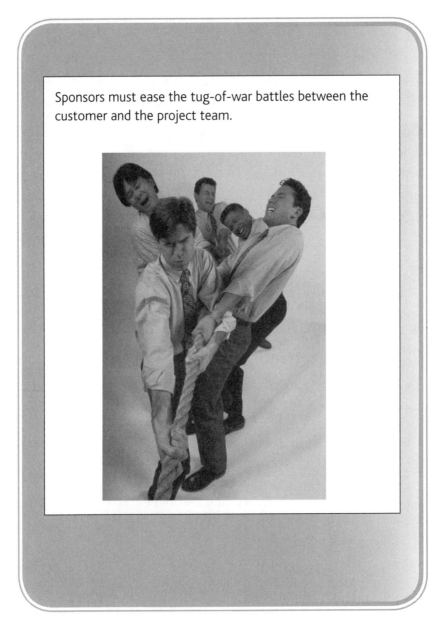

No matter how well statements of work (SOW) are written, disagreements and misunderstandings will occur after the go-ahead decision is made. This can result in time-consuming tug-of-war battles between the customer and the project team. The impact on the elements of the triple constraint can be enormous.

Allowing team members to get caught in the middle of disagreements is generally considered poor management and is a bad idea. This can lead to frustration, poor morale, and additional conflict. Sponsors generally have an easier time smoothing and working through disagreements. As stated previously, sponsors may act as referees during conflicts and may be required to protect the team during conflict resolution.

Sponsors must encourage short lines of communication. This can be difficult on multinational projects.

THE EXECUTIVE SPONSOR'S ROLE

Companies have finally recognized the high cost of face-to-face communications. Meetings are expensive when airfare, ground travel, meals and lodging are included. Another often forgotten issue is the potential for nonproductive time associated with traveling to and from these meetings and the time spent away from work.

Sponsors must encourage short lines of communication. This can be difficult on multinational projects where time zone differences of 10 to 12 hours may exist. Using available technology for virtual meetings and conferences over the Internet is one commonly applied solution.

It is often better for the project sponsor, rather than the project manager, to placate an irate customer.

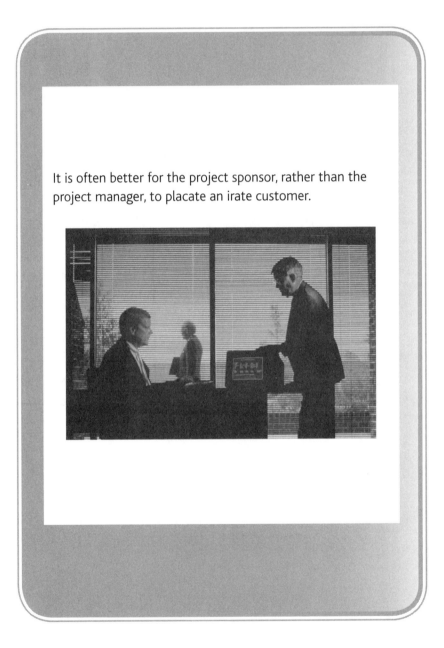

Customers have a tendency to overreact when bad news appears. Sometimes the overreaction is the result of other work in the customer's organization that is highly dependent on the completion of the project in question.

It is often better for the project sponsor, rather than the project manager, to placate an irate customer. Customers seem to respond more positively when a senior manager becomes involved.

Customers are less likely to act erratically and emotionally when communicating with the sponsor as opposed to the project manager. Also, sponsors generally possess the authority to make immediate decisions to resolve the issues. Sponsor involvement leads the customer to believe that senior management is involved in and overseeing the project.

The executive sponsor must make sure that a recognition or reward system is in place, even if it is a nonmonetary reward system.

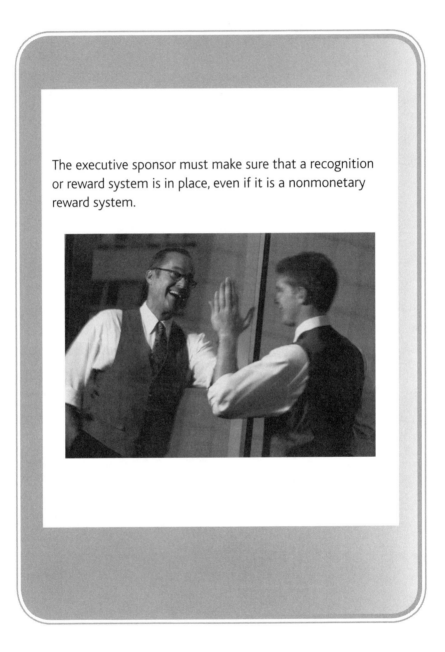

When employees are first assigned to project teams, their first and foremost concern is, "What is in it for me?" Team members want to believe that, if the project is successful, there will be some personal value involved and rewards will be forthcoming. Unfortunately, it is difficult to set up a reward system that satisfies the needs of all project employees and addresses those who do not work on project teams.

Critical questions to be addressed include:

- Who decides on the type and size of the reward?

- Who decides on each person's contribution to the success?

- How will nonproject employees be affected?

- How will this reward system affect future projects?

Some companies are focusing more on establishing a nonmonetary reward system. This includes:

- Achievement plaques and certificates

- Tickets to popular events—sports, concerts, and the like

- Gift certificates for dinner or performances

- Use of a company car

- Complimentary time off

Some executive sponsors believe that, based on the customer, the single most important project management skill is the ability to deliver a presentation.

Project managers generally have sufficient technical knowledge concerning the project to present data to the customer during interface meetings. Project managers may be expected to make all or part of the presentation.

An often overlooked skill that project managers need is the ability to deliver a presentation. Poor presentation delivery skills could leave the customer with concerns as to how the project is being managed. Some sponsors believe that presentation skills should be part of all project management training programs. When customers see a well-prepared presentation, they are more likely to feel confident that the project is managed the same way. A poorly prepared presentation could have serious implications for the project manager and the performing organization.

Sponsors must realize that projects are managed by people, not computers or software.

Project management is a people-oriented process. People are needed to manage projects and will never be completely replaced by computers or software.

- People control computers and software, not vice versa.

- Computers and software are and always will be tools for the project team to use.

With the low cost of computers today and the abundance of available software, project management has become heavily dependent on available software. But as a retired Air Force lieutenant general stated, "Never let the tool control the hand that's holding it."

All project sponsors, irrespective of their level in the hierarchy, must maintain an open-door policy.

Timing is of the essence in project management. Procrastination and/or other delays in decision making can have a serious impact on the project's schedule. Also, customers are generally impatient and unforgiving concerning delays in decision making.

Many project decisions ultimately fall on the shoulders of the project sponsor. As such, the project sponsor must maintain an open-door policy. The open-door policy should be in place irrespective of the sponsor's position in the organizational hierarchy. Also, the open-door policy should exist for all team members and not exclusively for the project manager.

There is a risk, however, when announcing that the sponsor maintains an open-door policy for project-related issues. People sometimes go directly to the sponsor when, in fact, the issue should first go through the appropriate chain of command, depending, of course, on the nature of the issue. Line managers react very negatively when they hear about issues from the sponsor first rather than from their own team members. Team members must understand that going to the sponsor is a last resort rather than the first step in the process, and the team members must know what type of issues to escalate to the project sponsor.

6

SPECIAL
PROBLEMS
FACING
EXECUTIVES

PUSHING SPONSORSHIP DOWN

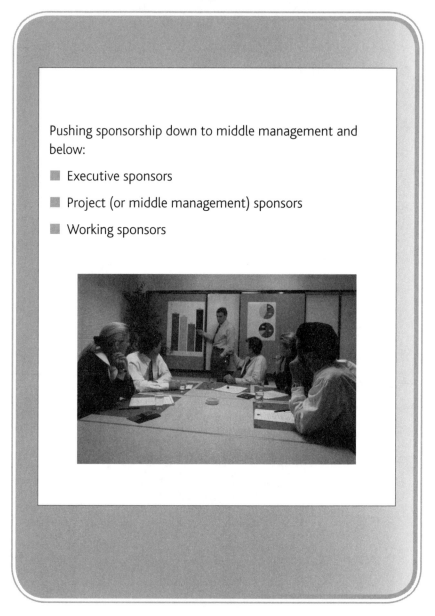

Pushing sponsorship down to middle management and below:

- Executive sponsors
- Project (or middle management) sponsors
- Working sponsors

A s shown previously, project sponsorship need not always be associated with senior levels of management. Some projects can be
sponsored effectively at lower or middle levels of management. Sometimes project managers prefer having sponsors at middle-management
levels because they are easier to access in times of trouble.

Some companies have defined multiple sponsors, at different levels
of management, for the same project. As an example:

- *Working sponsor.* Your immediate supervisor who assists the
  project manager on a daily basis

- *Middle-management sponsor.* Involved in the approval of scope
  changes

- *Executive sponsor.* Providing funding for the project and for
  scope changes.

Generally, one—and only one—sponsor is considered better than
having multiple sponsors. The multisponsor arrangement could
impact decisions and create unnecessary bureaucracy. On the next
page, we will discuss the criteria for deciding which level of management should function as the sponsor.

Deciding which level of management should sponsor certain projects:

- Strategic importance

- Expected profits

- Overall risk

- Technical knowledge

- Daily involvement

- Project management maturity

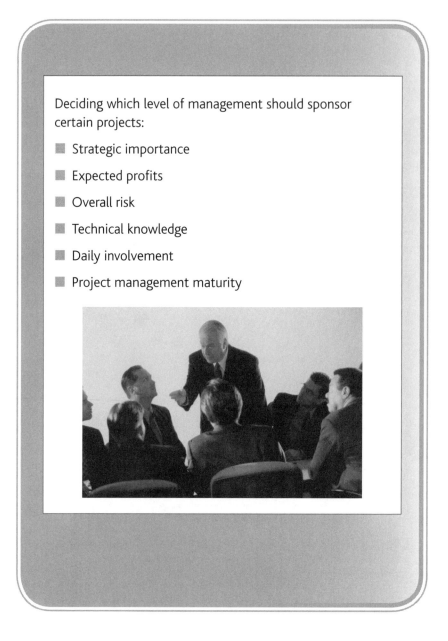

Companies that handle a multitude of projects but have relatively few executives have realized that the effectiveness of the sponsor is diluted when each executive must sponsor several projects at once while performing his or her normal line functions. Pushing sponsorship down the hierarchy is essential, especially as project management matures within the company. Factors that are often considered when deciding on the appropriate level include:

- How important the project is to the company

- Who the ultimate customer is and the impact on customer relations

- The project's target profit

- The amount of technical knowledge the sponsor needs for decision making

- Whether the sponsor needs to be involved on a daily basis

- The risk associated with the project

- The interrelationship between this project and other projects within the portfolio of projects

- The company's maturity level with regard to project management

COMMITTEE SPONSORSHIP

Some companies believe that sponsorship can be done through a committee. Critical questions include:

- How many people on a committee?
- Who should be the key sponsor?

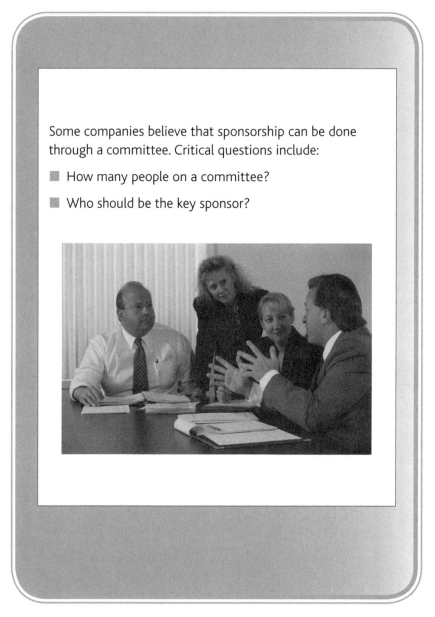

Having the same executive functioning as a sponsor through all of the life-cycle phases can alienate other line executives. For example, the vice president for marketing may be assigned as the sponsor but may not be qualified to make decisions regarding the engineering or research and development (R&D) life-cycle phases. Likewise, the vice president of R&D is unlikely to allow the vice president of marketing to make R&D decisions.

A common solution to this problem would be to have sponsorship by committee. While this approach seems plausible and has worked successfully in companies such as General Motors, there are two critical concerns:

- The number of people on the committee should be held to a minimum in order ensure productive meetings, streamline the decision process, and be able to meet frequently if problems occur.

- A committee of two or three usually works well.

- Should one person on the committee be appointed as the chair or "key" sponsor to manage disagreement within the committee?

HANDLING DISAGREEMENTS WITH THE SPONSOR

Project managers may not always agree with the decisions made by the sponsor. Policies and procedures can be established to resolve the disagreements.

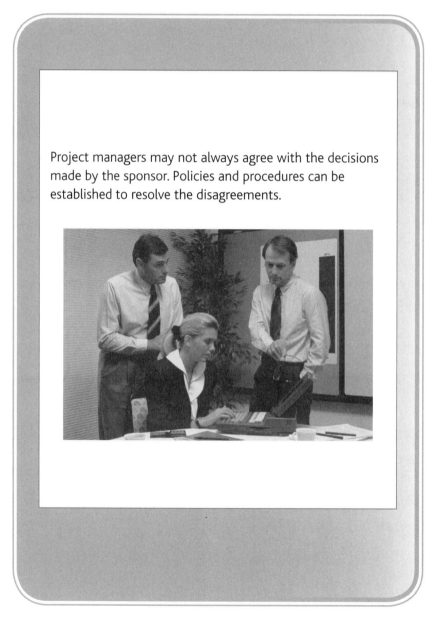

In the early years of project management, no one ever disagreed with the decisions made by the sponsor. But as project management matured and project managers assumed some authority and developed decision-making skills, it became evident that the project manager's input often resulted in better alternatives and decisions.

We know that the project sponsor and the project manager may not always be in agreement. Companies are now establishing executive steering committees and executive policy boards to handle these disagreements. When an issue occurs and the project manager and sponsor cannot come to an agreement, elevating the issue to a senior steering committee may work well.

The difficulty with this approach is that the sponsor may very well be a member of the steering committee. In this case, the sponsor must wear only the hat of the sponsor and cannot vote in favor of himself/herself as a member of the steering committee. While it seems on the surface that the members of the committee may favor the sponsor because he or she is also a member of the committee, this is not always the case. These committees seem to make decisions in the best interest of the company, as well as the best interest of the project.

KNOWING WHEN TO SEEK OUT THE PROJECT SPONSOR FOR HELP

Project sponsorship should not be viewed as a dumping ground for problems that the team does not wish to resolve. Problems should be brought to the sponsor:

- After all apparent options have been considered and are deemed unacceptable

- When it is obvious that the problem can be resolved only at the sponsor's level

However, when you bring an issue to the sponsor:

- Bring the problem as well as recommended options and alternatives to the project sponsor or there may be a risk that the sponsor will make a decision that will not be in the best interest of the project.

- Recognize that after the sponsor provides you with guidance, the problem remains the responsibility of the project manager, not the sponsor.

While some sponsors make decisions, many sponsors simply provide guidance. Sponsors do not generally have the time to actively investigate all issues brought to them. By bringing alternatives and recommendations, you are more likely to obtain support and effective guidance than by going in without any useful information.

TYPES OF SPONSOR INVOLVEMENT

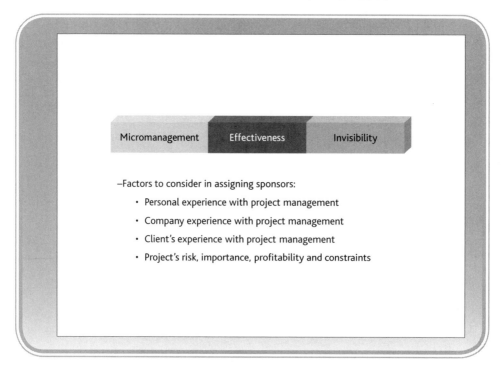

There are three types of project sponsor involvement:

- *Micromanagement.* This is excessive involvement in the day-to-day activities of the project. This occurs when the organization is immature in project management or when the sponsor is unsure about the project manager's capability.

- *Invisibility.* This is a situation where the project manager cannot get access to the sponsor and, even if access is granted, the sponsor refuses to make a decision or delays decision making. This often occurs when the executive is disinterested in the project or fears that a bad decision could impact their political career in the firm.

- *Effectiveness.* This is when the sponsor becomes involved on an as-needed basis.

The type of involvement selected by the sponsor is based on his or her faith in the team, the project's risks, the sponsor's experience with project management, the organization's culture, and the project management maturity level.

PLACATING THE (EXTERNAL) CUSTOMERS

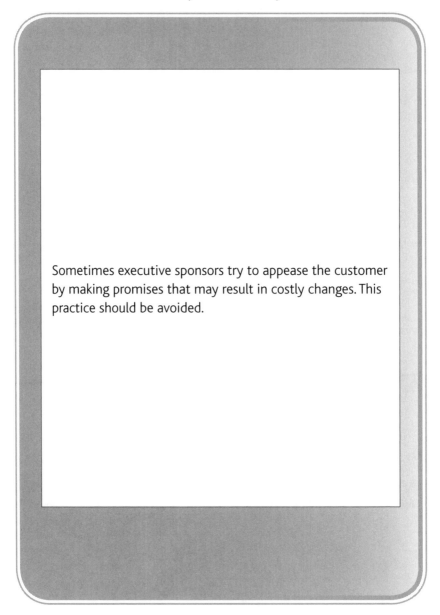

Sometimes executive sponsors try to appease the customer by making promises that may result in costly changes. This practice should be avoided.

Previously, we stated that sponsors may be the best people to interface with the customers when a crisis appears. When this happens, sponsors must try to placate the customer by making decisions that satisfy both the performing company and the customer. Unfortunately, sponsors can do a great deal of damage to the project without fully recognizing it.

Inexperience may cause sponsors to make promises to the customer such that cost changes will result. The sponsor may not have the ability to add additional funding. Sponsors must understand the risks involved with making promises about out-of-scope work to the customer. Customers often appear irate and threatening in an attempt to get the sponsors to make promises that result in little or no additional cost to the client. Examples include:

- "We'll run three additional tests to verify the results."

- "Let's have a few additional interface meetings and I'll bring the project team to you."

- "We'll provide you with a few additional reports describing our solution to the problems."

Each of these statements incurs significant expenses for the project.

GATE REVIEW MEETINGS

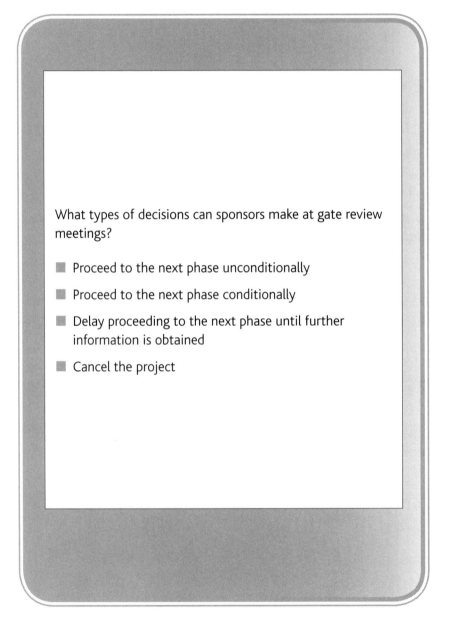

What types of decisions can sponsors make at gate review meetings?

■ Proceed to the next phase unconditionally

■ Proceed to the next phase conditionally

■ Delay proceeding to the next phase until further information is obtained

■ Cancel the project

One of the benefits of using life-cycle phases is that end-of-phase reviews may be scheduled or gates are established that the project must pass through to determine if the project should continue on to the next phase. The keys for opening the gates rest in the hands of the sponsors, not the project managers. This allows sponsors to be involved in critical decision points rather than having to meddle throughout the project.

In the illustration we see the four types of decisions a sponsor can make at the gate review meetings. The most difficult decision is the cancellation of the project. The reason why this is so difficult is that significant funds may have already been spent and the cancellation could appear to be a waste of company resources and may look bad for the sponsor, especially if the sponsor arranged for the financing.

Gate review meetings are usually held at the end of a life-cycle phase. But some projects that require rapid creation of deliverables may allow the life-cycle phases to overlap. In such a case, the gate review meetings may be held periodically throughout the project rather than at the end of a life-cycle phase.

SPONSORSHIP PROBLEMS

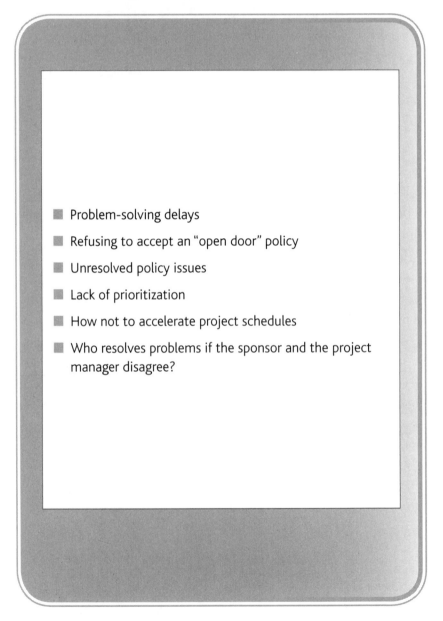

- Problem-solving delays
- Refusing to accept an "open door" policy
- Unresolved policy issues
- Lack of prioritization
- How not to accelerate project schedules
- Who resolves problems if the sponsor and the project manager disagree?

These problems are a composite of several issues previously discussed. The project sponsor position may be perceived as one in which a project sponsor seems like a glorified position where one simply sits back and makes decisions. But, as can be seen, there are numerous problems that can make life difficult for project sponsors.

If a project fails, the risks to the sponsor can be greater than the risks to the project manager because the sponsor was most likely the source of funds for the project, if an internal project. The failure can drastically damage the sponsor's career goals and reputation.

These problems are created by both the company and the sponsor. Some problems are created by the company's policies and procedures that must be followed. Other problems are created by sponsors who do not understand their own role.

THE EXIT CHAMPION

- What is an exit champion?

    - Validation of the existence of "exit ramps"

- Who normally functions as an exit champion?

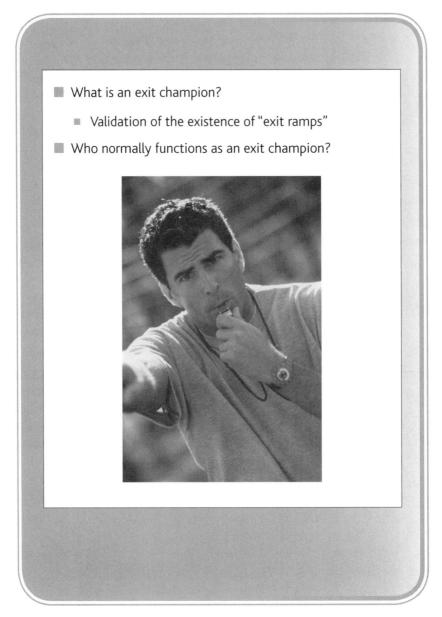

In addition to project sponsors or champions, some companies are creating the position of exit champion. Project sponsors are often reluctant to allow their project to be canceled for fear of damage to their career and reputation. Exit champions are most often executives who may have no personal interest in the project, financial or otherwise, and whose role is to periodically evaluate the project from an organizational perspective.

Projects must have exit ramps in case it becomes obvious that the project must be canceled. Just as the name implies, exit champions focus on the termination, as early as possible, of non-value-added projects. Exit champions must be able to recognize the value or lack of value in the project. Champions also must have a good understanding of the industry, the business, and environmental factors that can influence the success of the project.

SHOULD A SPONSOR HAVE A VESTED INTEREST?

Vested interest or not?

❖ Vested interest
  – Finance the fund-starved project
  – Keep project alive
  – Maximum protection from obstacles
  – Fend off internal enemies
  – Actively involved
  – Involved in personnel assignments

❖ Impartial
  – Provide no funding and limited support
  – Let project die
  – Limited protection from obstacles
  – Avoid politics and enemies
  – Go through motions
  – Partial involvement in assignment

Some people believe that the sponsor should be assigned from the organization funding the project, and therefore the sponsor will have a vested interest in seeing the project completed successfully. Others argue that a vested interest makes it difficult to make a decision to "pull the plug" on a project that is not achieving its desired value, so resources can be reassigned to more beneficial projects. Assigning an exit champion in addition to a project sponsor can alleviate this issue.

PROJECT CHAMPIONS VERSUS EXIT CHAMPIONS

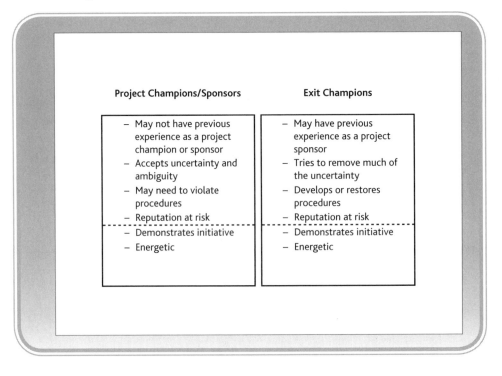

Project Champions/Sponsors	Exit Champions
– May not have previous experience as a project champion or sponsor – Accepts uncertainty and ambiguity – May need to violate procedures – Reputation at risk – Demonstrates initiative – Energetic	– May have previous experience as a project sponsor – Tries to remove much of the uncertainty – Develops or restores procedures – Reputation at risk – Demonstrates initiative – Energetic

This illustration compares project sponsors with exit champions. Some activities are the same, as shown below the dotted line in the figure. However, above the dotted line we can see the differences in their activities.

There is merit in having an exit champion who has had previous experience as a project sponsor. There may also be some merit in assigning exit champions who have had previous project management experience in order to determine whether the project can be saved.

THE COLLECTIVE BELIEF

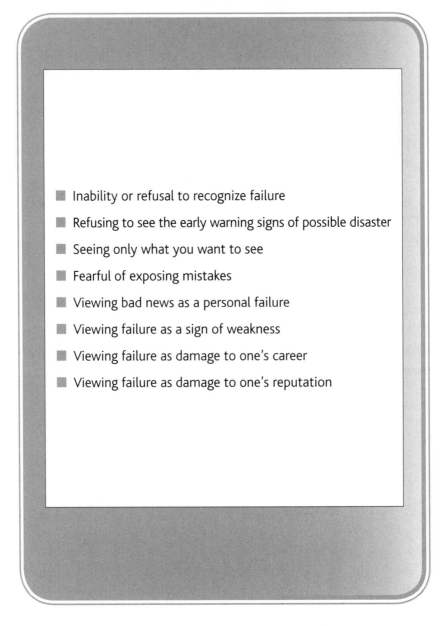

- Inability or refusal to recognize failure
- Refusing to see the early warning signs of possible disaster
- Seeing only what you want to see
- Fearful of exposing mistakes
- Viewing bad news as a personal failure
- Viewing failure as a sign of weakness
- Viewing failure as damage to one's career
- Viewing failure as damage to one's reputation

One of the reasons for assigning an exit champion is that large, long-term projects may have developed a "collective belief." This collective belief is a fervent, and perhaps blind, desire to achieve, which permeates through the team up to and including the project sponsor. The belief can make a rational organization act irrationally. The belief does not support a well-defined review process and, even if a review process and a review exist, the belief will impact the objectiveness of the review and people are not allowed to challenge any negative results.

Failure is considered unacceptable to the collective belief. Team members and sponsors can demonstrate the characteristics shown in this illustration. Without an exit champion, the project may never come to an end.

ADVERTISING SPONSORSHIP

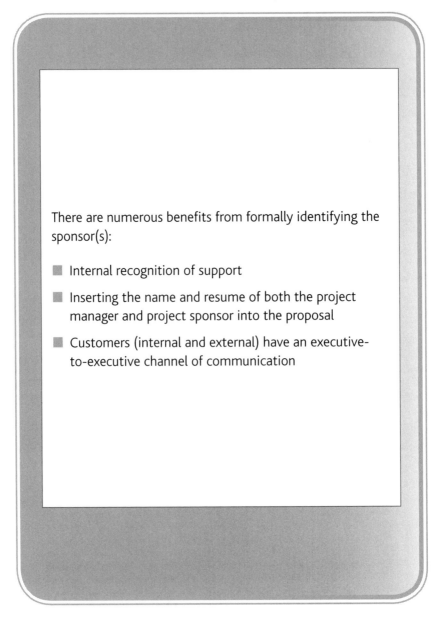

There are numerous benefits from formally identifying the sponsor(s):

■ Internal recognition of support

■ Inserting the name and resume of both the project manager and project sponsor into the proposal

■ Customers (internal and external) have an executive-to-executive channel of communication

It has become common practice for companies that survive on competitive bidding to include in their proposal the name and resume of the project manager. His or her information is included in order to emphasize the company's experience with project management and commitment to project success.

One company placed in its proposal the names and resumes of two executives who could be assigned as the sponsor, and the customer was allowed to select which sponsor they preferred to work with. This practice can lead to better customer relationships and increased customer satisfaction.

WORKING WITH THE ON-SITE REPRESENTATIVES

Some companies may request that the contractor allow on-site representatives to be assigned for the duration of the project. This can be beneficial if it accelerates the decision-making process, provides for rapid decision making, and provides the contractor with some of the customer's strategic information.

The downside risk of having on-site representatives is that they usually reside close to or even within the contractor's project office. Having team meetings without involving the onsite representative may be difficult, especially when proprietary information is being discussed. Also, the on-site representatives may believe they have the authority to go wherever they wish in the company. The project sponsor must work with the project manager and establish a policy regarding what the on-site representatives can and cannot do. This is commonly referred to as the *rules of engagement*.

KICKOFF MEETING FOR PROJECTS

It is a good practice for executives, especially the project sponsors, to show their support by appearing at the initial kickoff meeting for the project.

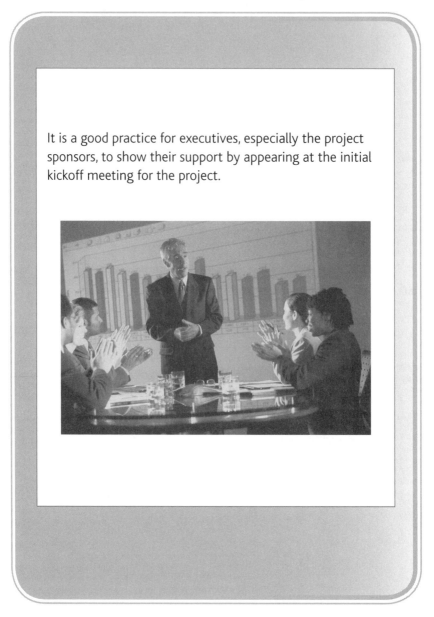

Previously, we stated that the project sponsor should be present at the initial project kickoff meeting to clearly articulate the project's objectives and connection to organizational goals. But there are other reasons, including:

- To provide an opportunity for the team meet the project sponsor

- To communicate clearly to the team that the company endorses the project

- To identify the aspirations of the sponsor and other executives

- To identify all project assumptions made, especially corporate assumptions

- To discuss the expectations that the sponsor has of the team

- To discuss that the sponsor exists to support the project manager and the team

- To explain the role of the sponsor as it relates to each life-cycle phase

TAKING THE LEAD

- General endorsement to start the enterprise project management (EPM) process
- Identify his/her own aspirations
- Identify corporate assumptions

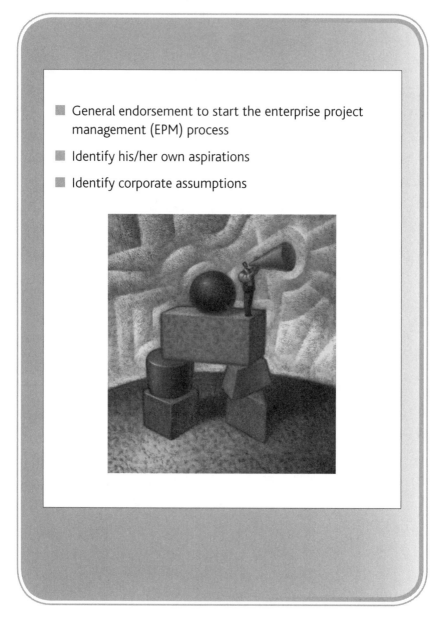

It is a good practice for project sponsors to make an appearance at the initial kickoff meeting to interface with the team. It is not a good practice for the sponsor to meet team members only when there is a crisis.

Although sponsors are not necessarily involved in the day-to-day activities, they are still part of the project team and should be recognized as a team member. Their immediate visibility in front of the team implies executive support for the project and executive interest. Team members need to believe that senior management has an interest in the project's outcome.

REWARDING PROJECT TEAMS

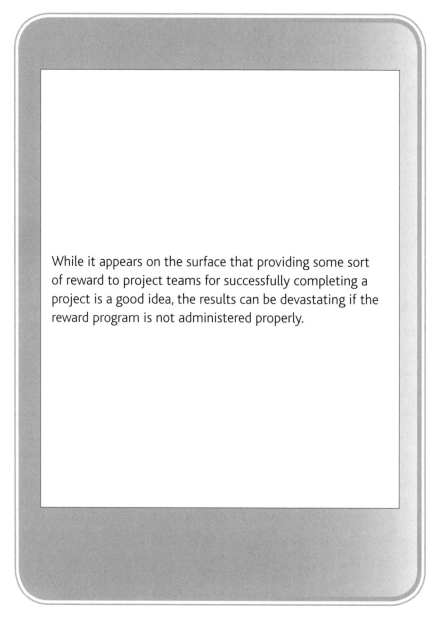

While it appears on the surface that providing some sort of reward to project teams for successfully completing a project is a good idea, the results can be devastating if the reward program is not administered properly.

The sponsor must make sure that the team fully understands any reward system that may be in place, and make sure it is equitable to all employees. If a reward system is in place, then the sponsor must discuss:

- The criteria for determining the overall reward

- The criteria for determining each individual's reward, whether the employee is assigned full time or part time

- How each person's contribution will be measured

- Who will determine each person's contribution

- The role of the line manager during the award process

ENTERPRISE PROJECT MANAGEMENT

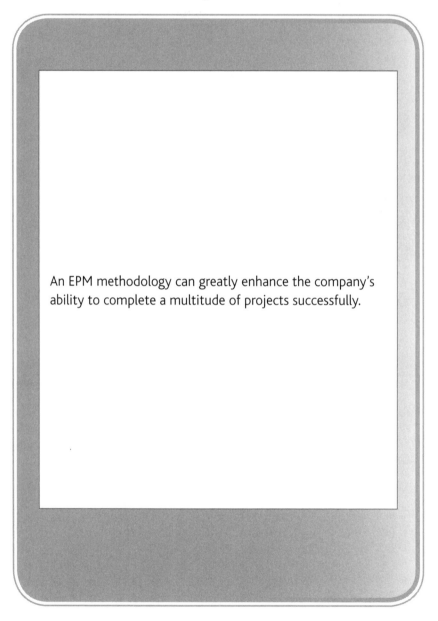

An EPM methodology can greatly enhance the company's ability to complete a multitude of projects successfully.

The sponsor must endorse the use of the company's EPM methodology. The methodology should be used on all projects and for all customers. However, it should be understood that not all parts or steps in the methodology may be required for every project. Many methodologies are scalable to meet the needs of the project.

There is a tendency within many organizations to establish criteria to determine when the methodology should be used. The criteria may be associated with the size of the project, the risks, the duration of the project, and/or the number of functional units that must interface during project implementation. The problem with this is that executives may have the authority to arbitrarily change the criteria to an extent where the methodology will not be used.

EXECUTIVE INVOLVEMENT (WITH TRADE-OFFS)

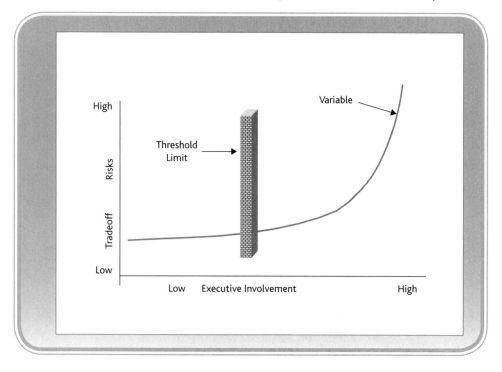

The most common trade-offs during project implementation projects are those associated with time, cost, performance, and risk. Project managers generally do not have the authority to make these trade-off decisions; the authority may reside with the sponsor.

Sponsors set threshold limits for themselves regarding what type of trade-offs they will be involved with and when they will get involved. As seen in the illustration, the critical threshold limit is usually risk. Executive involvement is minimal when the trade-off risks are low.

NEW CHALLENGES FACING SENIOR MANAGEMENT

MEASURING PROJECT MANAGEMENT SUCCESS AFTER IMPLEMENTATION

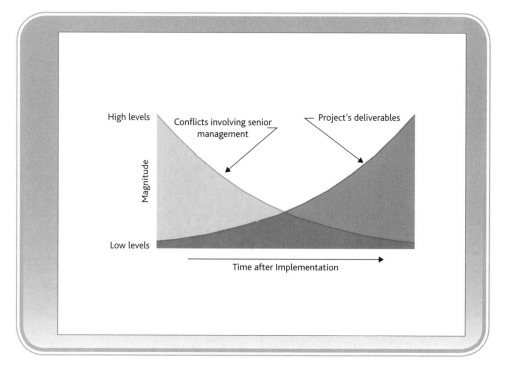

Once a project begins, sponsors become concerned as to whether project management is working correctly. There are two barometers that can be used:

- *Quantitative measurements.* Are we completing the deliverables within the project's constraints?

- *Qualitative measurements.* How many conflicts are being escalated to the project sponsor for resolution?

The qualitative measurement is critical. If fewer conflicts are coming to the sponsor for resolution, then it may appear that the project team and the line organizations are resolving the issues among themselves.

SUCCESS

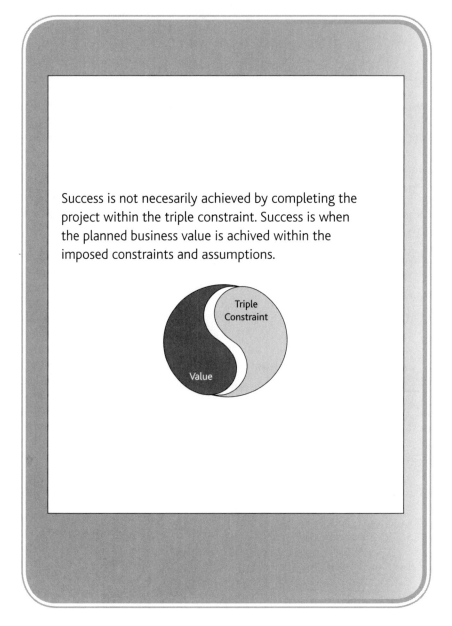

Success is not necesarily achieved by completing the project within the triple constraint. Success is when the planned business value is achived within the imposed constraints and assumptions.

Historically, sponsors focused on the definition of project *success* simply by looking at performance based on the triple constraint. While these factors are important, real success is achieved when the planned business value is achieved. This may imply that we may need to extend the project's duration or increase the cost baseline to guarantee that the business value will be realized. Completing a project within the triple constraint is no guarantee that value will be achieved at the completion of the project.

Projects are (or at least they should be) undertaken for the value they are expected to provide. For some projects, the value may not appear until some time later, perhaps years, after completion. On other projects, the value may be evident immediately. Both the customer and the contractor must define project value and recognize when it is achieved. They must also determine when the desired value will not be realized.

TYPES OF VALUES

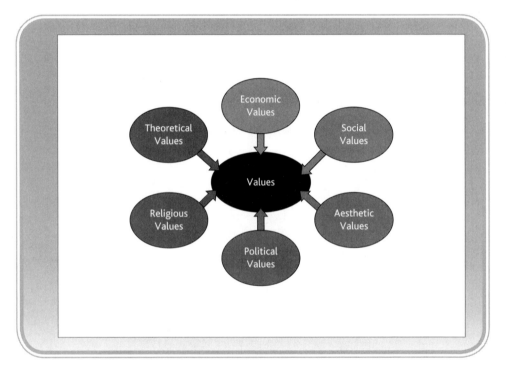

A s seen in the illustration to the left, there are many types of per-ceived value. Because professional responsibility is now an integral part of project management, we could argue that all these types of value apply to project management, including:

- Adhering to ethical standards

- The appropriateness of receiving or providing gifts

- Adhering to security and confidentiality requirements

- Truthfully reporting information

- Willingness to identify violations

- Recognizing the value of diversity in the project environment

- Managing customer/contractor intellectual property

- Maintaining professional integrity

- Understanding global, cultural, religious, and ethical beliefs

For this segment, project management value entails the attributes of your enterprise project management (EPM) system, appearing in the form of goods or services, for which your customers are willing to pay. The project using the EPM system must produce distinct products as services with features and characteristics that are valued by the customers.

FOUR CORNERSTONES OF SUCCESS

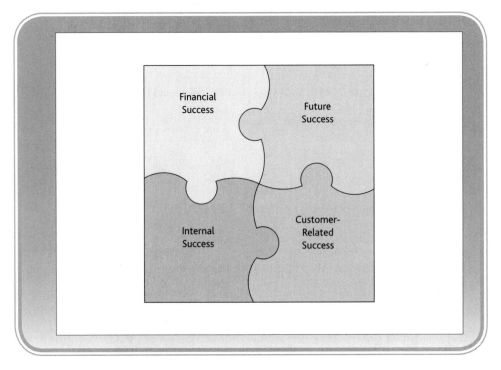

Defining project success has never been an easy task. The focus has always been the triple constraint. Today, we believe that there are four cornerstones for success:

- *Internal success.* The ability to have a continuous stream of successfully managed projects using an EPM methodology and that continuous improvement occurs on a regular basis.

- *Financial success.* The ability to create a long-term revenue stream that satisfied the financial needs of the stakeholders.

- *Future success.* The ability to produce a stream of deliverables that will support the future existence of the firm.

- *Customer-related success.* The ability to satisfy the needs of the customers over and over again to the point where you receive repeat business and the customers treat you as though you are a partner rather than a contractor or supplier.

SUCCESS VERSUS FAILURE

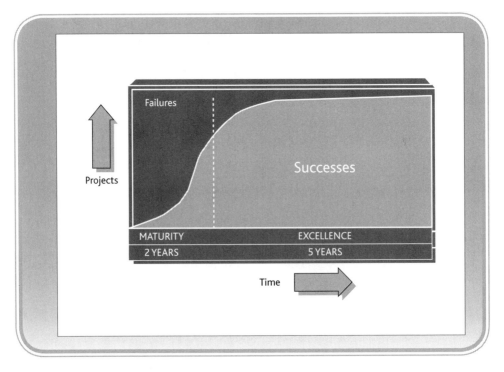

Successful implementation of project management takes about two years. To achieve true excellence in project management may take much longer. But even with excellence in project management, there is no guarantee that success will be forthcoming. Project management cannot guarantee that the project will be successful, but it can improve the chances for success.

The reasons are:

- A consistent methodology can improve the predictability of outcomes, but there is no guarantee the outcome will be the right outcome.

- A process for continuous improvement can be established, but the results may not be implemented in the near term.

- Guidelines for managing project issues and change will be in place, but too many issues and changes can have a serious impact on the triple constraint.

- The objectives of the project and the risks associated with the project may be clearly defined initially but may not be tracked appropriately, especially if enterprise environmental factors change.

HIGH-LEVEL PROGRESS REPORTING

High-level progress reporting needs to answer four fundamental questions:

- Where are we today (time and cost)?
- Where will we end up (time and cost)?
- What are the present and future risks?
- Are there special problems and what can management do to help?

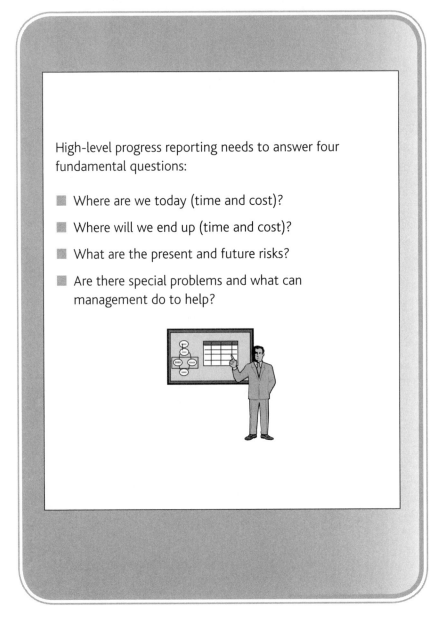

Executives and sponsors focus on the questions shown in the illustration on the left as part of high-level progress reporting. When project managers feel the necessity to present some of the more detailed information, it could become an invitation for executive-level micromanagement. Setting expectations about what information should be provided is an important part of the status reporting process.

It should be understood that if executives desire more detailed explanations, they will ask and the project manager should be prepared to provide the level of detail requested. It is a good practice for sponsors to clearly delineate to the project managers what information they would like to see in routine progress reporting. The information requested can change from project to project and sponsor to sponsor.

VALIDATING THE ASSUMPTIONS

Sponsors must take the time to revalidate the assumptions throughout the life cycle of the project.

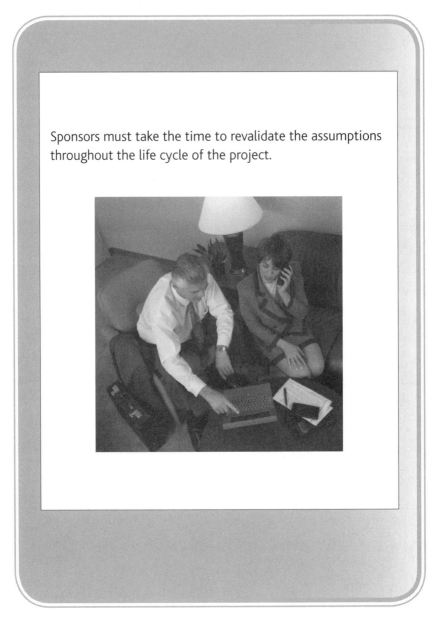

Once the project begins, there is a natural tendency to assume that the assumptions made at the beginning of the project are valid for the entire life cycle of the project. This is often not the case.

The assumptions must be periodically revalidated. The majority of the assumptions involve business conditions and the environment rather than technology. If the assumptions are no longer valid, then perhaps the direction of the project needs to be changed or possibly the project should be canceled. The revalidation must be performed routinely by the project sponsor and exit champion.

ACCELERATING PROJECTS

Any single project can have its schedule accelerated when executives exercise their formal authority and use a "big stick." But this can result in many of the other projects coming in late.

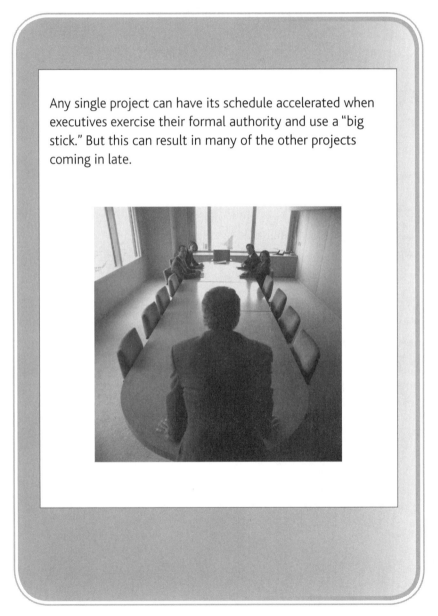

When projects experience schedule trouble, executives often believe that they can accelerate the schedule or resolve the issue by exercising their authority or by using a "big stick" approach. A common way of compressing a schedule is by adding more resources. The problem with this approach is determining where the added resources will come from, what additional training will be required for the newly assigned resources, who will provide the training, and the costs associated with the resources.

Taking resources from other projects to help accelerate the schedule on a project experiencing trouble may cause the other projects to be delayed, which may result in other, even more serious problems. The same result can occur when a lower-priority project suddenly becomes a high-priority project. Critical resources are then shared or moved, and some projects will inevitably suffer. Companies generally do not have resources standing by idly waiting for an assignment. Capacity planning allows maximum usage of the resources. Shifting the resources arbitrarily can change the capacity profile of the company.

PROJECT MANAGER SELECTION

Executives are responsible for selecting a project manager to head up the team.

Selection of the project is, for most organizations, an executive decision. The project sponsor may also have the greatest influence regarding who will be assigned to the project. Factors that can influence the sponsor's decision include:

- Business knowledge requirements

- Technical knowledge requirements

- Project management experience

- Leadership skills

- Team-building skills

- Previous experience with specialized customer requirements

- Skills at integration management

- Reputation

- The need for feasibility studies and economic analyses

DELEGATION OF AUTHORITY

Executives must provide the project manager with sufficient authority with regard to decision making.

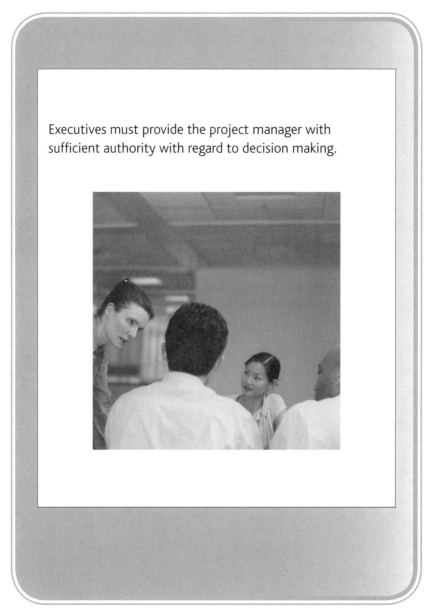

Sponsors do not have the time to review project issues and make all of the decisions. They must provide project managers with sufficient authority to make decisions associated with daily project demands and issues. The source and level of the authority can be described in the project's charter, which is signed by the sponsor. Factors that sponsors consider in providing authority include:

- Type and complexity of decisions to be made
- The project's risks
- The project manager's business and technical knowledge
- The project manager's previous experience in project management
- The project manager's previous experience with this client
- The size and nature of the project
- The company's maturity in project management
- The customer's experience in project management

VISIBLE SUPPORT

Executives must visibly demonstrate enthusiasm for and commitment to the project, the team, and project management.

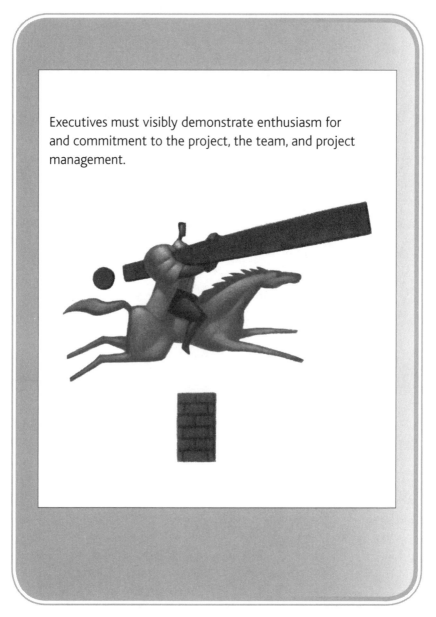

Previously, we stated that project sponsors should, whenever possible, attend the project's kickoff meeting to endorse the project and show executive support. This enthusiasm and commitment should be evident throughout the project life cycle rather than just at initiation.

Sponsors must be visible. This can be accomplished by attending a team meeting occasionally or by using a walk-the-halls management approach. Team morale improves significantly when the team sees executive involvement and has the opportunity to interface with the project sponsor.

CHANNELS OF COMMUNICATION

Executives must develop and maintain short and informal channels of communication.

As stated previously, the project sponsor position exists for the entire team, not exclusively for the project manager. As such, the sponsor must develop and maintain short, effective formal and informal channels of communication.

Every project team member should be given an opportunity to meet with the sponsor to discuss project-related issues. It is highly recommended that the sponsor maintain an open-door policy. The sponsor should arbitrarily force team members to go through the traditional chain of command in order to get access to the sponsor. Some type of informal channel or communications opportunity should be available. This type of communication can actually provide the sponsor with information that is more useful than information provided through formal channels and reports.

AVOID BUY-INS

Executives must avoid placing excessive pressure on the project managers to win contracts.

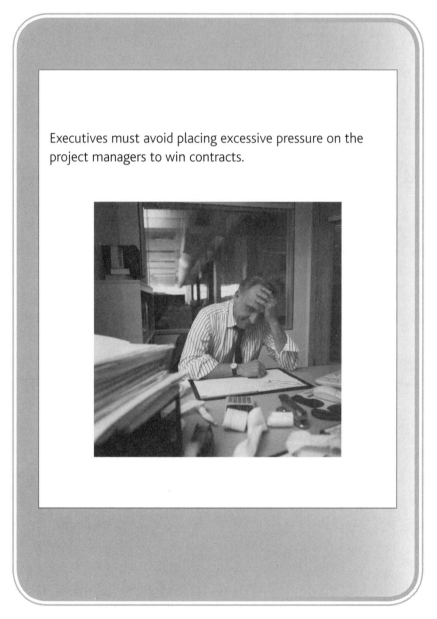

In some industries, project managers are actively involved in working with marketing and sales during competitive bidding activities. Project managers may even become responsible for writing the majority of the proposal. Executives must avoid placing excessive pressure on project managers to win contracts. Excessive pressure can lead to:

- Accepting unrealistic or unwanted risks

- Taking shortcuts that can lead to disaster

- Making promises that cannot be fulfilled

- Risking damage to the company's brand, image, and reputation

- Making it difficult to obtain follow-on work

- Conflict between project manager and functional managers regarding how to assign critical resources

BUDGETING

Executives must avoid arbitrarily slashing or ballooning the project team's cost estimate.

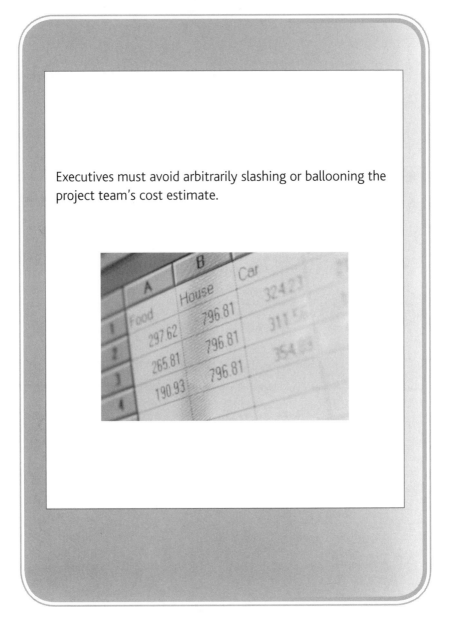

Some executives believe that they can arbitrarily slash or balloon the project's cost estimate. This can occur during competitive bidding prior to actual bid submittal, or even after go-ahead. It is unrealistic to slash the project's cost estimate without understanding the basis for the estimate and expect all of the work to be completed without experiencing severe project risks.

Ballooning the cost estimate can be equally as bad. Ballooning a budget normally occurs when scope is added to the project. If the added scope is unrealistic or cannot be rationally justified, more harm than good can result. Also, ballooning a budget and adding more work implies that additional resources will be available and can be assigned to the project. This can lead to disaster if all of the resources are already committed to other projects.

WORKING RELATIONSHIPS

Executives must develop close, not meddling, working relationships with the principal client contact and the project manager.

The sponsor must develop a close, but not meddling, working relationship with the client and the project team. Excessive meddling with the project team can easily lead to micromanagement and damage project team morale.

The more dangerous situation occurs when the sponsor tries to get too close to the principal client contact. Sometimes sponsors antagonize clients by seeking out information regarding:

- The client's strategy plan

- Future business opportunities with the client

- Possible value-added scope changes

While there may be some justification for the sponsor to do this, it must occur without alienating the client.

Chapter

8

ADDITIONAL RESPONSIBILITIES FOR EXECUTIVES

THE NEW ROLE FOR EXECUTIVES

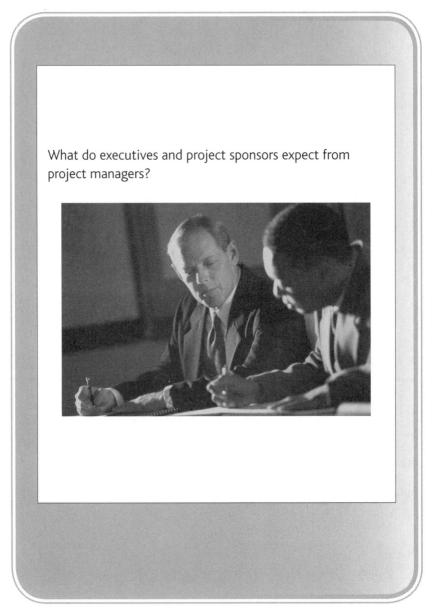

What do executives and project sponsors expect from project managers?

Executives are generally concerned more with end results than with processes. But if the processes do not undergo continuous improvements, then there may be concern about what happens to project deliverables over the long term. Without capturing lessons learned and best practices, organizations can become stagnant or inefficient. The changing project environment requires an organization to continuously assess abilities and develop new processes in order to sustain competitive advantage.

One popular solution is for management to establish a project officer (PO) or project management office (PMO). The charter for the PMO basically positions the PMO as the guardian of all project management intellectual property, and assigns responsibility to make sure that the processes in enterprise project management (EPM) methodology undergo continuous improvements. This includes capturing lessons learned and best practices from completed projects, interviewing subject matter experts, and conducting external benchmarking.

Establishing project management as a career path position.

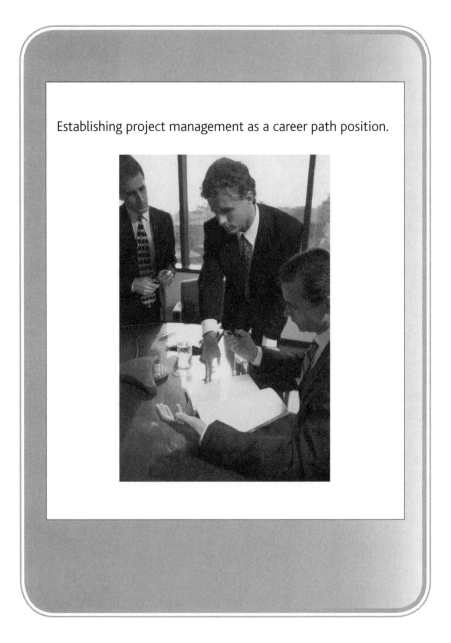

Senior management generally has the final say in establishing a career path for project managers. The project manager career path should provide the employees with the same opportunities and incentives for advancement as they would have with any other career path. An unequal or biased career-pathing arrangement within an organization may result in avoidance by employees to engage in a specific discipline.

The PMO is often given the responsibility to work with executives in the development of a project management career path.

Creating a mentorship program for newly assigned project managers.

If project management is expected to grow and mature, there must be a mentorship program created for newly assigned or inexperienced project managers. Since executives may not have the time to perform this mentorship, the work is often delegated to the PMO as part of their day-to-day activities. The mentorship program can also function as a hot line for project managers experiencing problems and can accelerate the development of solutions. Mentorships often result in improved project manager performance and may create new mentors for other aspiring project managers.

Establishing an organization committed to benchmarking best practices in project management.

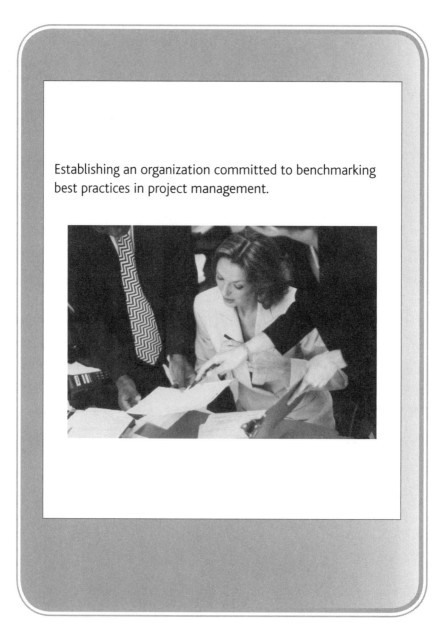

Improving project management capabilities requires a process that will capture knowledge that is internal or external to the organization. External knowledge comes primarily from benchmarking. There are three common forms of benchmarking:

- Process benchmarking

- Industry benchmarking

- World-class benchmarking

World-class benchmarking generally gives the best results because it provides more comprehensive information about the best practices associated with other companies in an industry and can be used for analyzing strategic decisions. Benchmarking activities are usually placed under the control of the PMO. One other important note about benchmarking is that it should not be restricted to one's own industry. Benchmarking against companies with excellent reputations for project management can provide fruitful results.

ACTIVITIES FOR A PROJECT MANAGEMENT OFFICE

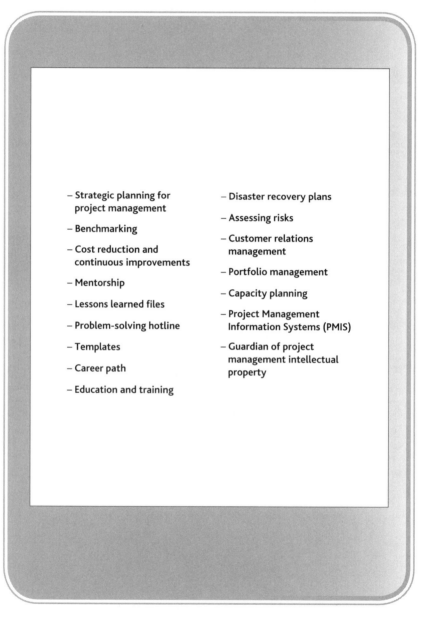

- Strategic planning for project management
- Benchmarking
- Cost reduction and continuous improvements
- Mentorship
- Lessons learned files
- Problem-solving hotline
- Templates
- Career path
- Education and training

- Disaster recovery plans
- Assessing risks
- Customer relations management
- Portfolio management
- Capacity planning
- Project Management Information Systems (PMIS)
- Guardian of project management intellectual property

There are numerous activities that executives can assign to a PMO. These are shown in the illustration on the left. A typical PMO will not deliver projects for clients, but will manage projects related to the continuous improvement of project management methodologies. The PMO can also develop processes that will make the role of the project sponsor easier to perform.

THE EXECUTIVE INTERFACE

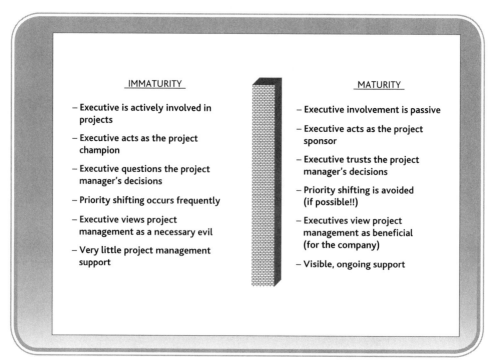

IMMATURITY

- Executive is actively involved in projects
- Executive acts as the project champion
- Executive questions the project manager's decisions
- Priority shifting occurs frequently
- Executive views project management as a necessary evil
- Very little project management support

MATURITY

- Executive involvement is passive
- Executive acts as the project sponsor
- Executive trusts the project manager's decisions
- Priority shifting is avoided (if possible!!)
- Executives view project management as beneficial (for the company)
- Visible, ongoing support

This illustration shows the differences in how executives interface with projects in a mature project management organization versus companies that are immature.

In a mature environment, the sponsor is positioned as a support resource, emphasizing trust in the project team, a "hands-off" approach to daily activities, and clearly supporting the benefits and values of a project management methodology.

EXPECTATIONS

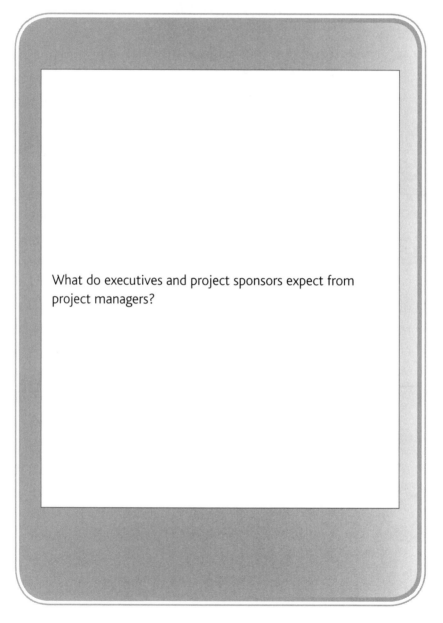

What do executives and project sponsors expect from project managers?

Executives and sponsors have certain expectations of project managers. These include:

- Assuming total accountability for the success or failure of the deliverables

- Providing complete and accurate reports and information

- Minimizing organizational disruption during the execution of a project

- Presenting recommendations, not just alternatives

- Managing most project-related interpersonal problems

- Demonstrating a self-starting attitude and capability

- Demonstrating growth with each new assignment

What do project managers expect from project sponsors and executives?

Project managers have expectations of executives and sponsors, including:

- Providing clearly defined decision channels

- Taking action on requests and recommendations

- Facilitating the interfacing between and with support departments

- Providing assistance during conflict resolution

- Providing sufficient resources

- Providing constructive feedback

- Providing protection from political infighting

- Providing opportunities for growth

A STRUCTURED PATH TO MATURITY

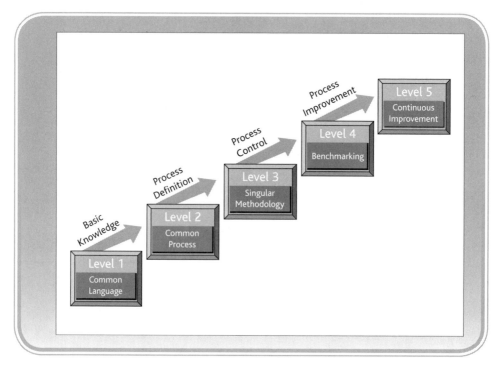

Achieving maturity in project management should not be left to chance. There are structured and unstructured paths to achieve maturity. A typical structured path is the Project Management Maturity Model (PMMM) shown on the left.

- *Level 1.* The organization recognizes the importance of project management and a need for basic understanding.

- *Level 2.* The organization recognizes the importance of creating repeatable processes.

- *Level 3.* The repeatable processes are combined into an EPM methodology.

- *Level 4.* The organization recognizes the importance of continuous improvements through benchmarking.

- *Level 5.* The organization evaluates the benchmarking information to see what can be implemented.

AN UNSTRUCTURED PATH TO MATURITY

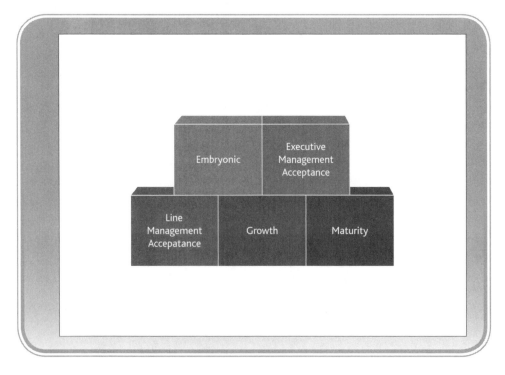

This illustration shows an unstructured path to maturity:

- *Embryonic phase.* Recognize the needs, benefits, and applications of project management.

- *Executive Management Acceptance phase.* Recognize the need for visible executive support and sponsorship.

- *Line Management Acceptance phase.* This includes educating the line managers and then getting their support and commitment.

- *Growth phase.* This is the development of a project management methodology.

- *Maturity phase.* This includes the development of a cost/schedule control system and project management skills training for the workers.

CONCLUSIONS

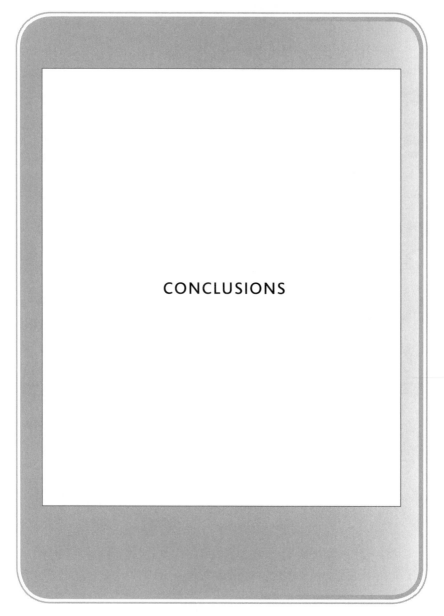

An executive understanding of project management is essential for long-term growth and maturity in project management. Executives must provide proper sponsorship for projects and should not set expectations that are beyond the capabilities of the team.

Executives possess the authority and power to grow and improve project management capabilities or destroy the support for project management and eliminate the processes. Once the executives understand the true benefits of project management, plans are established and steps are taken to achieve some level of maturity in project management.

Possible additional conclusions:

- Project management is beneficial to an organization.

- Project managers must be given some level of authority to achieve project objectives.

- Project sponsors should manage with a "hands-off" approach to daily project activities.

- Project sponsor visibility is key to project success.

- A project management career path will improve the quality of project manager performance.

- Project success is not limited to the triple constraint.

- Setting expectations between project sponsor and project manager is critical.

- The role of the project sponsor should be clearly defined and communicated at the project kickoff.

- The project sponsor role goes beyond project funding and includes mentorship, liaison with other executives and functional groups, and the administration of reward and recognition programs.

- The project manager provides key insight to project selection and project portfolio management.

The basic principles of project management associated with planning, executing, and controlling have remained relatively constant through the ages. These principles have driven the definition of success that is associated with the triple constraint.

The value of project management, however, continues to evolve and the perception of achieved value also continues to evolve. We have clearly shown that the expected project value and the actual value achieved may be quite different and the true value of a project may not be realized until many years after the project has been completed. Executive managers are encouraged to observe very closely the key indicators of success that are currently in place within the organization and consider reviewing and possibly revising the perception of project value and how value is measured in the near term and the long term.

Additionally, connecting with project managers occasionally through informal meetings, as well as structured executive briefings, will provide an opportunity to obtain a "front line" view that will create a more positive relationship between executive and project manager and provide the executive with a more realistic view of, and a greater appreciation for, project management.

INDEX